NATIONAL DEFENSE RESEARCH INSTITUTE

T0146158

National Guard Youth ChalleNGe

Program Progress in 2017–2018

Louay Constant, Jennie W. Wenger, Linda Cottrell,
Wing Yi Chan, Kathryn Edwards

Prepared for the Office of the Secretary of Defense
Approved for public release; distribution unlimited

For more information on this publication, visit www.rand.org/t/RR2907

Library of Congress Cataloging-in-Publication Data is available for this publication.
ISBN: 978-1-9774-0250-9

Published by the RAND Corporation, Santa Monica, Calif.

© Copyright 2019 RAND Corporation

RAND® is a registered trademark.

Cover: *U.S. Air National Guard photo by Master Sgt. Becky Vanshur.*

Support RAND
Make a tax-deductible charitable contribution at
www.rand.org/giving/contribute

www.rand.org

Preface

The National Guard Youth ChalleNGe program is a residential, quasi-military program for youth ages 16 to 18 who are experiencing difficulty in traditional high school. This report covers the program years 2017–2018 and is the third in a series of annual reports that RAND researchers will issue over the course of a research project spanning September 2016 to June 2020. The first and second National Guard Youth ChalleNGe Annual Reports cover program years 2015–2016 and 2016–2017, respectively, and can be found on the RAND website (Wenger et al., 2017; Wenger, Constant, and Cottrell, 2018). A fourth annual report, which will cover program years 2018–2019, will be released early in fiscal year 2020.

Each annual report documents the progress of participants who entered ChalleNGe during specific program years and then completed the program. A focus of RAND's ongoing analysis of the ChalleNGe program is collecting data in a consistent manner. Based on these data, each report also includes trend analyses. In the present report, we provide information in support of the National Guard Youth ChalleNGe Program's required annual report to Congress. In addition to information on participants who entered the ChalleNGe program and completed it in 2017, we include follow-up information on those who entered the program and completed it in 2016, as well as an update on our work to develop metrics to improve program effectiveness. Finally, we describe other ongoing research efforts to support the ChalleNGe program (these efforts will be detailed in future reports). Methods used in this study include site visits, collection and analysis of quantitative and qualitative data, literature reviews, and development of tools to assist in improving all program metrics—for example, a program logic model. Caveats to be considered include documented inconsistencies in reported data across sites, our focus on those who completed the program and not all participants, and the short-run nature of many of the metrics reported here.

This report will be of interest to ChalleNGe program staff, personnel providing oversight for the program, and policymakers and researchers concerned with designing effective youth programs or determining appropriate metrics by which to track progress in youth programs.

This research was sponsored by the Office of the Assistant Secretary of Defense for Manpower and Reserve Affairs and conducted within the Forces and Resources

Policy Center of the RAND National Defense Research Institute, a federally funded research and development center sponsored by the Office of the Secretary of Defense, the Joint Staff, the Unified Combatant Commands, the Navy, the Marine Corps, the defense agencies, and the defense Intelligence Community. For more information on the RAND Forces and Resources Policy Center, see http://www.rand.org/nsrd/ndri/centers/frp.html or contact the director (contact information is provided on the webpage).

Contents

Figures

Tables

Summary

The National Guard Youth ChalleNGe program is a residential, quasi-military program for youth ages 16 to 18 who are experiencing academic difficulties and exhibiting problem behaviors, either inside or outside school; have dropped out or are in jeopardy of dropping out; and, in some cases, have had run-ins with the law. The ChalleNGe program runs for a total of 17.5 months, broken into a 5.5-month Residential Phase (comprising a two-week acclimation period, called Pre-ChalleNGe, and the 5-month ChalleNGe) followed by a 12-month Post-Residential Phase. Participating states operate the program, which began in the mid-1990s, with supporting federal funds and oversight from state National Guard organizations. There are currently 39 sites in 28 states, the District of Columbia, and Puerto Rico. More than 220,000 young people have taken part in the ChalleNGe program, and nearly 165,000 have completed the program.

ChalleNGe's stated mission is to "intervene in and reclaim the lives of 16–18-year-old high school dropouts, producing program graduates with the values, life skills, education, and self-discipline necessary to succeed as productive citizens" (National Guard Youth ChalleNGe, 2010, p. 7). The program delivers a yearly, congressionally mandated report documenting progress; the data and information reported in this RAND report support this mandated report.

The ChalleNGe program emphasizes the development of eight core components: leadership and followership, responsible citizenship, service to community, life-coping skills, physical fitness, health and hygiene, job skills, and academic excellence. There is variation across the 39 ChalleNGe sites in the range of program activities implemented to support the program's core components. The factors that determine this variation are a combination of state and local context, program history, and site leadership preference. For example, although all sites provide volunteer opportunities for ChalleNGe participants (or cadets, as they are called) to fulfill the component of providing service to the community, the types of opportunities differ based on the nature and range of partnerships that individual sites have developed. There is also variation across sites in terms of what specific academic credentials are emphasized or offered. For instance, many cadets work toward a General Education Development (GED) or High School Equivalency Test (HiSET) certificate, but some sites allow cadets to work toward a high school diploma or collect high school credits that allow them to transfer back to and graduate from their home high schools after returning home. Although sites tend to emphasize one approach over another, many sites offer multiple options for cadets, and a cadet's choice typically will depend on his or her age, credit status at the time of enrollment, and personal or family preference. Additionally, some sites offer specific occupational training or certificates, and some offer the opportunity to earn college credits. Previous research has found that the ChalleNGe program is cost-effective and has positive effects on cadets' lives: ChalleNGe participants achieve more education and have higher earnings than similar young people who do not attend the program (see Bloom, Gardenhire-Crooks, and Mandsager, 2009; Millenky, Bloom, and Dillon, 2010; Millenky et al., 2011; and Perez-Arce et al., 2012). However, the ways in which site-level variations might result in different levels of long-term effectiveness are not known.

Objectives of the Project

RAND's ongoing analysis of the ChalleNGe program has two primary objectives. The first is to gather and analyze existing data from each ChalleNGe site on an annual basis to support the program's yearly report to Congress. This RAND report is the written product of the research team's third round of data collection. As in our first and second annual reports, we requested information from each ChalleNGe site. The core metrics collected have remained more or less consistent across the years. The bulk of the analysis in each annual report to Congress is based on these data. We collect measures that include tallies of the total number of young people who participated in and successfully completed the program, the number and type of credentials awarded, standardized test scores, participation in community service, and registration for voting and selective service. Each annual report also includes detailed site-specific information on participants, staffing, and funding. Finally, each annual report includes information on post-ChalleNGe placement (for example, enrollment in school, participation in the labor market, or enlistment in the military). We present measures at the site level; in some cases, we include aggregated information as well.

The second objective of this project is to identify longer-term metrics for the overall effectiveness of the program, including ones that will help determine how site-level differences influence program effectiveness. The relevant information comes from multiple sources, including measures collected for the annual report, analysis of qualitative data collected from site visits, and analysis of extant data, such as national samples matching the profile of the population of interest. In this report, we continue to document our progress toward this objective, and we share preliminary findings on two of these efforts: developing benchmarks for at-risk youth and examining the mentoring component of the ChalleNGe program.

Cross-Site Metrics for the 2017 ChalleNGe Classes

The quantitative information included in this report was collected from ChalleNGe sites in fall 2018. The RAND study team developed an Excel spreadsheet template to record program-level and cadet-level information. This template, first developed in 2016, was revised in 2017 and again in 2018 with each round of data collection. Changes include improvements and refinements to the indicators, as well as additional questions informed by site visits; the fundamental data elements collected have remained the same across years. The spreadsheet template was distributed to all sites during each of the three rounds of data collection.

It is important to note that schedules vary somewhat across ChalleNGe sites; we requested that each site send information on the cadets who entered the program in 2017. Typically, there are two cohorts of cadets in a year per site, with notable exceptions. The two classes described in this report are referred to by program staff as Class 48 and Class 49 (ChalleNGe classes are numbered consecutively from the first class in the 1990s).

Our current findings reveal that Classes 48 and 49 produced nearly 10,000 ChalleNGe graduates. Around 64 percent of graduates received at least one academic credential. As a group, these graduates performed more than $14 million worth of service to their communities. The overall graduation rate for these two classes was 73 percent, based on the number of cadets who enter the two-week acclimation period termed Pre-ChalleNGe.

To date, we have collected information on six classes of ChalleNGe cadets; our previous reports include information on Classes 44 to 47 (Wenger et al., 2017; Wenger, Constant, and Cottrell, 2018). As a result, we can examine trends over time. Our trend analysis indicates that ChalleNGe programs took in more participants in 2017 than in 2016. Some of this increase is due to data from sites that did not report last year—e.g., in California, Georgia, and Tennessee—but some is due to existing programs enrolling more cadets. The overall graduation rate, however, remained roughly constant with prior classes, as did the proportion of graduates who scored at or above the ninth-grade level on the standardized test used to track cadet progress. In sum, more young people graduated from ChalleNGe in 2017 than in 2016, and academic quality appears to be roughly constant over the study period.

Tests of Adult Basic Education Scores

The study examined Tests of Adult Basic Education (TABE) scores to ascertain changes in cadet progress between the beginning and just before completion of the program, as well as overall progress across programs and across time. The analysis revealed substantial progress, with three-quarters of cadets scoring at the eighth-grade level or lower at the beginning of the Residential Phase compared with more than half of cadets scoring at or above the ninth-grade level by the end of the Residential Phase. Scores on the TABE reading test were higher than scores on the math tests at both the beginning and end of the Residential Phase.

The study also noted variation across program sites that may influence the number of applicants, the graduation rate, or the academic progress cadets make in their programs. The contexts in which programs operate differ, including in the curriculum and services available in schools, the number and type of alternative programs available in the state, the extent to which ChalleNGe is well known in the state, and the background and experience of the instructors and the cadre who oversee the cadets. An important variation across programs is which credential the program predominately awards. Based on the information collected from the sites in fall 2018, programs were divided into two groups: programs primarily granting GED or HiSET certificates and those primarily granting high school credits or diplomas.

Examining differences in TABE scores by type of credential awarded, we found that graduates at sites granting high school credits or diplomas are slightly more likely to score at the elementary or middle school level when entering the ChalleNGe program. Scores for cadets in both types of programs are very similar at the end of the Residential Phase, although cadets at sites granting high school credits or diplomas are still more likely to score at the elementary or middle school level and less likely to score at the upper high school level. These TABE score results suggest that cadets entering programs granting GED or HiSET certificates are, on average, slightly better prepared for coursework than cadets entering programs that grant high school credits or diplomas. Although graduates of programs granting high school credits or diplomas make substantial progress—and have higher test score gains than graduates of GED or HiSET programs—even at graduation there is no indication that cadets at programs granting high school credits or diplomas hold an advantage in terms of test scores. In terms of graduation, cadets who enter programs that grant high school credits or diplomas are slightly more likely to graduate than cadets in GED or HiSET programs.

As a first step toward determining the relationship between program-level differences and cadet success, the study team collected information on several program-related policies, including admissions factors, disciplinary policies, policies related to placement of cadets in platoons, additional activities provided to cadets, job placement supports, and college placement supports. Most sites incorporated behavioral or mental health considerations, as well as juvenile justice records and physical health, in their admissions considerations. Sites also reported using a variety of policies to encourage positive behavior, including the use of usual privileges and through Corrective Action through Physical Exercise (CAPE) to enforce disciplinary policies and the promise of additional privileges on campus to cadets who follow rules and policies. Sites typically do not place cadets from the same family in the same platoon, and around half of sites reported taking a cadet's hometown, friendships, and gang membership into consideration.

In terms of activities and ceremonies, most sites offered student government, a graduation ceremony, a color guard, a library, and an awards banquet. Most sites provided a range of activities to assist cadets with job placement after completing the ChalleNGe program, including mock interviews, assistance with job seeking and completing job applications, career tests, and various other career exploration activities. Most sites also help cadets with college placement supports, including helping cadets complete applications and providing access to relevant information. The study team will continue to explore potential relationships between program-level activities and cadet success.

Staffing

The study team also collected information on the number of staff by position, as well as the number of staff by position newly hired in the past 12 months. At about half of all sites, at least 25 percent of total staff were hired in the last 12 months. Among the older programs (established prior to 2000), a relatively large share of recruiters were newly hired within the past 12 months. Among newer programs, at least half reported that 25 percent or more of their instructors, case managers, and cadre were newly hired in the past 12 months. In general, staff turnover at the most established programs was lower than turnover at newer programs, and turnover among all programs tends to be highest among cadre, followed by recruiters.

Program Completion

Of youth who enroll in the ChalleNGe program, roughly 25 percent leave prior to graduation. In this report, the analysis largely focuses on cadets who complete the program; however, we present descriptive statistics to examine factors that are associated with graduation. Consistent with past work, female cadets graduated at a higher rate than male cadets and, to a lesser extent, cadets entering at age 17 or 18 seem to be more likely to complete the program and graduate than those entering at 16 years of age. Graduation rates also tended to be slightly higher at programs awarding high school credits or diplomas, and graduation rates were slightly lower at programs with a higher reported cadre turnover in the past year.

Placement

The study team also examined placement data collected and reported by the program sites. By months six and 12, at least three-quarters of graduates are listed as having a placement, with education and employment as the most common placements. Cadets are most likely to be enrolled in school in the first month after graduation; in later months, cadets are more likely to be employed. The proportion of cadets who report military service increases over the months, as does the proportion who report some combination of education, employment, and military service. Notably, programs struggle to obtain placement information on all cadets. Among Classes 48 and 49, sites reported information on 80 percent of cadets at the six-month mark and 73 percent at the 12-month mark.

Time Trends, 2015–2017

With three years of data reported by the program sites, it is possible to show time trends on some key indicators, including total number of participants, graduates, and cadets whose TABE total battery scores exceed the ninth-grade level at graduation. Trends reveal that the number of cadets participating in ChalleNGe increased slightly from 2015–2017, as did the number of graduates, though the graduation rate remained roughly constant. The number of cadets scoring at or above the ninth-grade level at graduation also increased, although this growth was very slight and was much smaller than the overall growth in graduates. The proportion scoring at or above the ninth-grade level on the TABE remained roughly constant. Overall, the number of participants and the number of graduates increased without a marked decrease in graduation rates or achievement, as measured by test scores.

Additional Analytic Efforts

In addition to preparing this year's annual report, the third in the series, the RAND study team also undertook several analytic efforts that address components of the National Guard Youth ChalleNGe program and relate to the program's logic model. The logic model describes the design of the National Guard Youth ChalleNGe program, the activities that different program sites implement, and the expected outputs and outcomes. Our additional analytic efforts are intended to address gaps in data collection, particularly around long-term outcomes, and better understand program design and implementation issues (for instance, how to improve the mentoring component).

These analytic efforts draw on multiple data sources—including information collected annually from the program sites, qualitative data collected from site visits, and other extant data sources, such as publicly available data from nationally representative surveys—and they were developed in parallel with program site visits, during which the RAND research team identified key issues through interviews with site personnel. The underlying research questions were developed based on an assessment of the salience of the issues revealed in the interviews, team member expertise, and budget and timeline constraints.

In this report, we review a few of the analytic efforts that we developed over the past year in support of the ChalleNGe program. The first set of analytic efforts, which draws on best practices from the broader literature, includes research on a variety of topics relevant to the program's eight core components, particularly mentorship, career and technical education (CTE), and community service. A second set of analytic efforts seeks to employ empirical methods to analyze data relevant to ChalleNGe. For example, using publicly available data from nationally representative surveys, the RAND research team is developing benchmarks to allow a rough comparison between ChalleNGe cadets' outcomes and the outcomes of similar young people who do not take part in the program. Another empirical effort is designed to determine the regional payoff of specific credentials; this information will be useful as the program considers which credentials should be offered and emphasized. An additional analytic effort focuses on ChalleNGe graduation rates and how program-level differences may be reflected in performance differences across sites. In Chapter Three, we share preliminary insights from two specific analytic efforts: on benchmarks and on the mentoring component.

The aim of our benchmarking analysis was to develop realistic goals and expectations for ChalleNGe participants based on how young people who fit the same profile perform on key education, labor market, and psychosocial outcomes. Currently, programs have limited access to information that allows them to compare their graduates—and their outcomes—with a benchmark group; benchmarks will help sites judge the effectiveness of their efforts. The aim of our mentoring analysis was to address challenges that programs face in promoting robust and enduring relationships between mentors and cadet mentees. Our site visits and the data collected for this annual report reveal that mentor and mentee interaction typically drops off significantly within a few months and is particularly difficult to maintain after cadets have completed the program. We looked to the literature to identify approaches to improve the resilience and sustainability of the relationship between mentors and mentees.

The ChalleNGe program logic model—which we introduced in the first annual report (Wenger et al., 2017)—illustrates the design features of the intervention and delineates the inputs, processes and activities, expected outputs, and desired outcomes of the program. The logic model provides a basis for identifying metrics that determine progress toward the achievement of program goals. Currently, most sites focus on collecting and reporting on metrics associated with the inputs, activities, and outputs. Metrics on short-term outcomes (achieved within one year) are collected and reported on, but the extent and consistency of reporting varies from site to site.

Benchmarking Study

To develop benchmarks, we used nationally representative surveys of young adults—including the National Longitudinal Survey of Youth (NLSY), the Current Population Survey, and the American Community Survey—to compare three mutually exclusive populations: high school dropouts who do not attain a GED, high school dropouts who attain a GED, and high school graduates who do not attend college. These three populations are the most appropriate comparison for the youth served by the ChalleNGe program and thus are appropriate benchmarks to use for tracking outcomes, including degree completion, employment, and incarceration. These expected outcomes can help form expectations of what the programs can achieve and provide a rule-of-thumb benchmark for sites to measure their own success. We compare these three groups across an array of metrics and at different ages. For example, during the dropout window (ages 14–18), we compare the share of each group who reported a learning disability or got suspended while enrolled in high school; at ages 19–29, we compare the share who are regularly employed and their average earnings. From the three averages of each group, we develop a high-low range, or benchmark, for each outcome. In our analysis, the high and low estimates were often—but not always—set by dropouts and graduates, with GED holders falling in between. We propose that this range approximates reasonable high- and low-end expected outcomes for a high school dropout intervention program. We describe here the process we used to conduct the analysis, and we preview some of our key findings; the full findings from this analysis will be published in a separate report.

Mentoring Study

The ChalleNGe program typically employs a youth-initiated mentoring (YIM) approach to match mentors to cadets. That is, incoming cadets nominate an individual meeting certain criteria to be their mentor. In the minority of cases in which cadets do not nominate a mentor, one is assigned to them. For this analytic effort, we examined the literature on YIM, and we shared insights on the implementation of mentoring from our site visits, with a focus on YIM. There is emerging evidence to support the use of YIM. For example, a prior study of the ChalleNGe program found that cadets who chose their own mentors had longer relationships with them in comparison with cadets who were assigned mentors; this study also found that youth who had longer mentoring relationships reported better outcomes compared with counterparts who had shorter mentoring relationships (Schwartz et al., 2013). Another study of ChalleNGe found that mentees reported feeling supported by, and close to, their mentor. Mentees also noted that their mentors had helped them become more confident and more capable of having positive relationships with adults, as well as helping them make progress toward their post-ChalleNGe goals (Spencer et al., 2016).

To gain a better understanding of the implementation of the ChalleNGe mentoring program, we analyzed interview notes from 22 site visits completed in 2017 and

2018. Our qualitative analysis revealed barriers to implementing a robust and sustainable mentoring program. Efforts to improve the program could include more training for mentors, while better tracking of mentor and mentee interactions could shed light on the relationship between mentoring and key youth outcomes of interest.

Pilot Programs

In addition to the activities described above, the RAND research team is partnering with two ChalleNGe sites to initiate pilot programs to inform improvements to ChalleNGe programming in the areas of mentoring and post-residential placement tracking. The overall goal of the pilots is to identify improvements that all ChalleNGe programs can benefit from. For the mentoring pilot, the study team is working with a ChalleNGe site to strengthen the initial and ongoing training that mentors receive to better prepare and support them in their role. The pilot will be assessed with the aim of extracting recommendations that can be applied to improving mentoring across programs. Similarly, the study team is working with another program site to identify means of following and gathering data on cadets postgraduation using a simple survey to collect placement data, along with information on longer-term outcomes, such as education, employment, and life transitions. The efficacy of the tool and approach to collect data will be assessed by the study team, which will provide recommendations for improvement and information on the approach's efficacy in producing plausible information on cadet placement and outcomes postgraduation.

Conclusion

Data collected over the past three years indicate that cadets across the ChalleNGe program continue to make progress in many areas. The data also reveal that although all ChalleNGe sites carry out a core foundation of activities, there is considerable site-level variation in this regard. Further research, which the RAND team will undertake in advance of our next annual report, is needed to better understand how and to what extent this variation is correlated with certain outcomes of interest.

A key shortcoming of this examination of the ChalleNGe program is that information collected to date does not allow the measurement of longer-term outcomes and impacts. Developing such metrics will continue to be a primary focus of this project, and progress on this front will be documented in future reports. In addition to the benchmarking study and the examination of mentoring, the study team is conducting additional analytic studies in support of other topics relevant to ChalleNGe programming, including an examination of CTE opportunities available to cadets, community service activities provided by ChalleNGe, and the association between site-based characteristics and cadet outcomes. The overarching goal of this project is to help ChalleNGe sites track their progress and inform implementation of program improvements.

Acknowledgments

We are grateful to the staff of the National Guard Youth ChalleNGe program. We thank the central administrative staff who assisted with many aspects of this research and the staff at each location who responded to our data request in a timely fashion and also provided detailed and thoughtful information during the course of our site visits.

We are also grateful to our RAND colleagues for their support: Craig Bond and Sarah Meadows reviewed various portions of the report, Emily Payne provided administrative support, and Joy Moini managed project operations, including budget and timeline, and contributed as a member of the research team. We thank Libby May and Brian Dau, who edited the document for improved clarity and exposition at different stages. We also thank Beth Bernstein for managing the publications process. Gabriella Gonzalez of RAND and David DuBois of the University of Illinois at Chicago provided reviews to ensure that our work met RAND's high standards for quality. Jennifer Buck of Spectrum provided the graphic describing the logic model.

We thank all who contributed to this research or assisted with this report, but we retain full responsibility for the accuracy, objectivity, and analytical integrity of the work presented here.

Abbreviations

BMI	body mass index
CAPE	Corrective Action through Physical Exercise
CTE	career and technical education
GED	General Education Development
HiSET	High School Equivalency Test
NLSY	National Longitudinal Survey of Youth
P-RAP	Post-Residential Action Plan
RCT	randomized controlled trial
TABE	Tests of Adult Basic Education
TASC	Test Assessing Secondary Completion
TOC	theory of change
YIM	youth-initiated mentoring

CHAPTER ONE

Introduction

The National Guard Youth ChalleNGe program is a residential, quasi-military program for young people ages 16 to 18 who have left high school without a diploma or who are at risk of dropping out because they do not have enough credits to graduate given their age and associated grade level. ChalleNGe participants (or *cadets*) may be referred by school counselors or other school officials, law enforcement or the juvenile justice system, or other members of the community. Programs do require, however, that young people who participate do so voluntarily, and parents or guardians must consent to this participation.

Participating states operate the program through their state National Guard organizations with supporting federal funds and oversight. The National Guard is responsible for all day-to-day operational aspects of the program; the Office of the Secretary of Defense provides oversight. States are required by federal law to contribute at least 25 percent of the operating funds. The first ten ChalleNGe sites were established in the mid-1990s; today, there are 39 ChalleNGe sites in 28 states, the District of Columbia, and Puerto Rico.[1] More than 220,000 young people have participated in the ChalleNGe program, and roughly 165,000 have completed the program. Table A.1 in the Appendix includes a list of all ChalleNGe sites.

ChalleNGe's stated mission is "to intervene in and reclaim the lives of 16–18-year-old high school dropouts, producing program graduates with the values, life skills, education, and self-discipline necessary to succeed as productive citizens."[2] ChalleNGe is based on eight core components: leadership and followership, responsible citizenship, service to community, life-coping skills, physical fitness, health and hygiene, job skills, and academic excellence. ChalleNGe's overarching goal is to be

---

[1]   When we collected the data for this report, there were 40 sites operating; in late 2018, the two sites in Texas consolidated into a single site. We report data separately for the two sites throughout this document, but future documents will include information on only a single Texas site.

[2]   The mission statement can be found in previous annual reports to Congress (for example, National Guard Youth ChalleNGe, 2010, p. 7) as well as the ChalleNGe website (National Guard Youth ChalleNGe, undated). The mission statement appears to be widely shared across ChalleNGe sites. It is quoted in various materials and briefings used at the sites and was included in briefings that formed part of our site visits.

recognized as the nation's premier voluntary program for 16–18-year-olds who struggle in a traditional high school setting, serving all U.S. states and territories. Previous research has found that ChalleNGe has a positive influence on participants' near-term labor market outcomes (Bloom, Gardenhire-Crooks, and Mandsager, 2009; Millenky, Bloom, and Dillon, 2010; Millenky et al., 2011) and is cost-effective (Perez-Arce et al., 2012).[3]

RAND's ongoing analysis of the ChalleNGe program has two primary objectives. On an annual basis throughout the project, we collect and analyze data from each site in support of the program's yearly reports to Congress; this is the third such RAND report from this project.[4] The first and second reports included information on the ChalleNGe classes that entered the program in 2015 and 2016, respectively, as well as a description of the ChalleNGe logic model and analyses of Tests of Adult Basic Education (TABE) scores, body mass index (BMI), and mentor reporting. All three reports published to date aim to lay the foundation for developing longer-term metrics of cadet success. This report is designed as a stand-alone document; for this reason, it includes some common information with the previous reports. However, those reports (Wenger, Constant, and Cottrell, 2018; Wenger et al., 2017) include additional analytic details on the TABE, BMI, and the classes from 2015 and 2016.

Our second objective is to develop a rich and detailed set of metrics to capture more information about the long-term effectiveness of the program. To this end, we are undertaking a series of analytical efforts focusing on various aspects of the ChalleNGe program. In this report, we share preliminary findings from two of those analytic efforts, with more-detailed findings to be shared as a separate stand-alone report or in the project's final research report. The aim of the first analytic effort is to develop benchmarks for at-risk youth that will enable program sites to compare the education, employment, and social outcomes of their participants against youth in the broader population with the same profile. This analysis relied on publicly available data from nationally representative surveys to identify benchmarks. The methodological approach demonstrated in this analysis can be used to identify and quantify metrics in the logic model, specifically education, employment, and social outcomes.

The aim of the second analytic effort is to strengthen the mentoring component of the ChalleNGe program. This undertaking addresses commonly expressed concerns

---

[3]   MDRC researchers employed a randomized controlled trial (RCT) to evaluate the effects of ChalleNGe by comparing a treatment group (those who participated in ChalleNGe) with an otherwise similar control group who were not randomly assigned to participate in ChalleNGe. The researchers collected information using a survey at nine months, 21 months, and 36 months following entry into the study (Bloom et al., 2009; Millenky, Bloom, and Dillon, 2010; Millenky et al., 2011). The RAND study used the MDRC findings to conduct a cost-benefit analysis of the program, factoring in the projected lifetime earnings given higher educational attainment from participation in ChalleNGe (Perez-Arce et al., 2012).

[4]   This report draws heavily on information provided in our previous reports (Wenger, Constant, and Cottrell, 2018; Wenger et al., 2017).

across program sites about the difficulty of maintaining mentor engagement with mentees over time, particularly after graduation. Our analysis of the mentoring component looks to the literature and identifies best practices and approaches to support the development of more-robust youth-initiated mentoring. It also addresses a key component of the intervention design: the supports available to cadets in the program's 12-month Post-Residential Phase. We anticipate that improvements to the mentoring component will result in easier transitions for cadets from the program's 5.5-month Residential Phase to the Post-Residential Phase and will help ensure cadets' continued path to success in education, employment, and social outcomes. These benchmarks are intended to also help monitor program improvements.

The underlying framework for the entire study—and the connecting seam across the study's associated analytic efforts—is the program logic model. The logic model describes the design of the National Guard Youth ChalleNGe program, the various activities that program sites across states implement, and the expected outputs and outcomes. The data collected for the annual reports include much of the information related to inputs, activities, outputs, and short-term outcomes in the logic model. These additional analytic efforts are intended to address gaps in terms of data collection and particularly long-term outcomes and to better understand program design and implementation issues, such as improving the mentoring component. The end result of these and other analytic efforts will be (a) four annual reports, (b) shorter reports summarizing the findings of each analytic study, and (c) a final research report, structured around themes identified in site visits conducted from 2017–2019, that synthesizes the overall findings from RAND's ongoing study and that comprehensively addresses the components of the logic model and the ChalleNGe program.

In the remainder of this chapter, we provide additional background information on the ChalleNGe program. We then describe in more detail the focus of this report and the methodologies we used. We conclude with a road map for the remainder of the report.

The ChalleNGe Model

The ChalleNGe program has several distinctive characteristics. Because cadets generally attend a site located in the state where they live, not all young people have access to the program. Recruitment for the program varies from site to site. Typically, to ensure broad coverage, program sites conduct regular outreach to high schools (especially to counselors), organizations that run out-of-school programs serving young people, and other community-based health and education organizations that serve underprivileged youth and their families. Program representatives conduct on-site visits, give presentations about the program, and distribute marketing materials. High school counselors refer students to the program, but, in many cases, students or their parents reach out to the program directly after attending an informational event. Programs also rely on word of mouth from former graduates, family members, peers, and high-profile community members who support the program, especially in smaller and tight-knit communities. In some cases, young people are referred to the program by members of the juvenile justice system. Participation is voluntary, and there is no tuition cost to the cadet or his or her family. Cadets must apply to the program, however, and most sites have a "packing list" of items that cadets must bring on their first day. Many sites also require applicants to complete an interview or attend an information session at the site (or both). Most sites do not have minimum standardized test score requirements. Applicants must not be currently awaiting sentencing, on parole, or on probation for anything other than a juvenile offense; also, they must not be under indictment or accused or convicted of a felony (U.S. Department of Defense Instruction 1025.8, 2002).

The ChalleNGe program runs for a total of 17.5 months, broken into a 5.5-month Residential Phase and a 12-month Post-Residential Phase. During the Residential Phase, cadets reside at the site in a barrack-like atmosphere, wear uniforms, and perform activities generally associated with military training (e.g., marching, drills, physical training). The first two weeks of the program, referred to as the Acclimation Period, are designed to allow new cadets time to adjust to the environment, as well as to the expectations that the ChalleNGe program requires for success. Coursework begins at the end of the Acclimation Period. For the next five months—the main part of the Residential Phase—cadets attend classes for much of the day. The academic curriculum varies across sites; this variation is a result of program history, state context, and choices made by program leadership in each state.[5] Some sites focus on the completion of a GED or High School Equivalency Test (HiSET) credential. At other program sites, cadets have the option to earn high school credits that they can use to transfer to a high school at the end of the ChalleNGe program and go on to earn a high school diploma. Still other ChalleNGe sites award high school diplomas to cadets who complete the state requirements for high school graduation. Some sites give cadets the option to choose among these models.

Not all cadets complete the 5.5-month Residential Phase of the ChalleNGe program (completion is referred to as *graduation*). Most cadets who leave the program

---

[5]   Traditionally, the ChalleNGe program has focused on completing a General Education Development (GED) credential. However, some sites have transitioned to awarding credit recovery and high school diplomas. This transition has occurred as a result of several factors. First, leadership at some sites believe strongly that a high school diploma, rather than an equivalency degree, is needed to better ensure success in postsecondary education and the labor market, as well as eligibility for military enlistment. These sites have expended a concerted effort to make this transition. Second, some sites have attained affiliation with a local school district or the state as a charter school. This allows them to award high school diplomas or credits through recovery programs that are recognized by state and local authorities. Finally, sites are also responding to their local communities: in certain state and local contexts, parents and incoming cadets want to pursue a high school degree rather than an equivalency degree.

prior to graduation choose to withdraw, but sites can and do dismiss cadets who violate key policies. Cadets are not enlisted in the military during the Residential Phase, and there is no requirement of military service following completion of the program.

ChalleNGe places considerable focus on the development of noncognitive or socioemotional skills, such as having positive interpersonal relationships, developing goals and detailed plans to accomplish those goals, anger management, and attention to detail, among others. The program is based on the following eight core components:

- leadership and followership
- responsible citizenship
- service to community
- life-coping skills
- physical fitness
- health and hygiene
- job skills
- academic excellence.

Each ChalleNGe site is charged with developing cadets' skills and abilities in all eight areas. Mentorship plays a key role: each cadet has a mentor, and the relationship between cadet and mentor is intended to continue for at least 12 months after the cadet graduates from the Residential Phase (in other words, through the Post-Residential Phase). The ChalleNGe mentoring model is largely youth-initiated: cadets are encouraged to nominate their own mentors, and most do.

While there are no formal, professional qualifications for being a mentor, mentors must meet a set of criteria, including a minimum age. Mentors also must be the same gender as the cadet, be of good standing in the community, generally live in the same community as the cadet, not be an immediate family member of the cadet, and be willing to commit time to training and to attending regular meetings with the cadet. Mentors, who receive in-person training from ChalleNGe staff, are volunteers (i.e., they are not compensated). Mentors also maintain contact with program staff throughout the Post-Residential Phase. For cadets who have not identified an appropriate mentor when they enter the program, ChalleNGe staff will provide one.

The ChalleNGe model has been found to be effective through a randomized controlled trial (RCT) in two significant areas: youth who participate go on to complete more postsecondary education than youth who do not participate, and youth who attend the program are more likely to participate in the labor force when compared with similar young people who do not attend the program (Bloom, Gardenhire-Crooks, and Mandsager, 2009; Millenky, Bloom, and Dillon, 2010; Millenky et al., 2011). In a separate and careful analysis of the costs and benefits based on the outcomes from the RCT, RAND researchers found that ChalleNGe is also cost-effective, producing approximately $2.66 in benefits (appropriately discounted) for each $1.00 invested

(Perez-Arce et al., 2012).[6] The differences observed in the RCT included longer-term outcomes, such as GED attainment, traditional high school degree attainment, and college attendance, as well as employment and earnings up to three years after graduation; these outcomes are the reason the program was found to be cost-effective. These longer-term outcomes were collected specifically to conduct the RCT; sites do not regularly collect such information from their graduates. In many cases, the outcomes were self-reported in surveys administered as part of the study. This self-reporting could have influenced some outcomes; for example, crime- and health-related outcomes were found to be similar between youth who participated in ChalleNGe and youth who did not. An important limitation of the RCT and the RAND cost-benefit analysis based on the RCT results is that the positive effects of the program on youth were detected using only a subset of ChalleNGe sites. (For a more-detailed description of previous research on the ChalleNGe program, see Wenger et al., 2017.)

Focus of This Report and Methodology

This report, the third in a series of annual reports produced by RAND for this project, serves two purposes. The first is to *provide a snapshot of the ChalleNGe program on a range of quantitative indicators during 2017–2018. The second is to develop a set of metrics related to the long-term effects that ChalleNGe has on participants after they leave the program; these metrics will help document the extent to which the ChalleNGe program is achieving its mission.*

Given the twofold purpose of this report and our larger research agenda regarding the ChalleNGe program, we combined several methodologies: collecting quantitative data from each ChalleNGe site, developing tools to help determine the preferred outcome metrics from the ChalleNGe program, collecting qualitative data through site visits, and planning and beginning to carry out a series of other analytic efforts. We describe each of these efforts here and present additional detail in Chapter Three.

To provide a snapshot of the ChalleNGe program during 2017–2018, we include information gathered from individual ChalleNGe sites in summer to fall 2018. Much of this program-level information is typical of what was included in the first two annual reports. We collected and reviewed information from each site on program characteristics; 2017 budget and sources of funds; number of applicants, participants, and graduates; credentials awarded; and metrics of physical fitness and community service or engagement. We also collected information on staffing, the dates classes began and ended, and post-residential placements. We requested and received the informa-

---

[6]   Costs included the operating costs of the program as well as the opportunity costs of those participating. For more details about the RCT and the differences observed between ChalleNGe participants and similar young people who did not enter ChalleNGe, see Bloom, Gardenhire-Crooks, and Mandsager, 2009; Millenky, Bloom, and Dillon, 2010; and Millenky et al., 2011.

tion through secure data transfers (although we requested no identifying information). We specified that sites should include information from the two classes that began in 2017. This information meets the program's current annual reporting requirements and will be used in the program's 2018 report to Congress.[7] In Chapter Two, we provide program- and class-specific data and also provide some analysis of this information across programs.

As part of our data collection, we also requested cadet-level information on graduation, credentials awarded, changes in the TABE grade-equivalent scores, and placements during the Post-Residential Phase.[8] We also collected 12-month placement information from some 2016 classes; this information was not yet available at the time of our last data collection. Annual reports for the ChalleNGe program published prior to 2017 included only site-level measures and metrics,[9] such as the average gain in TABE grade-equivalent scores or the number of cadets placed; they did not include any cadet-level information.[10] Achieving *key levels* on the TABE predicts other relevant outcomes, such as passing the GED exam. Therefore, we used cadet-level information to report a series of metrics based on achieving key TABE levels (we developed these levels during previous years). We include some analyses of this information in Chapter Two; see Wenger, Constant, and Cottrell, 2018, and Wenger et al., 2017, for additional details.

To develop a set of metrics to assess longer-term impacts of the program, we developed a ChalleNGe-specific framework for defining longer-term outcomes. We began by developing two tools: a theory of change (TOC) model and a program logic model. These models serve as operational tools to guide the development of metrics and monitor progress toward achieving the program's central goals and evaluating its effectiveness. A TOC is a conceptualization of the mechanisms by which solutions can be developed to address a complex social problem; a program logic model delin-

---

[7]   See 32 U.S.C. 509(k) for annual reporting requirements.

[8]   TABE is currently developed by DRC/CTB, and its suite of tests is specifically designed to assess the basic skills of adult learners. According to the TABE website, workforce development programs in most U.S. states, whether funded or not funded by the federal Workforce Innovation and Opportunity Act, use TABE to assess the basic skills of individuals participating in their programs (TABE, undated). All ChalleNGe programs administer TABE to cadets at the beginning of the program and prior to graduation in order to measure academic achievement in math and language arts and to maintain a key metric by which to track cadet learning progress. TABE results are reported in past analyses; see, for example, the 2015 annual report (National Guard Youth ChalleNGe, 2015).

[9]   In technical terms, a *metric* is a specific value, while a *measure* refers to an activity, output, or outcome (National Research Council, 2011). Thus, the number of cadets who graduate from ChalleNGe could be considered a measure, while an overall cadet graduation rate of 80 percent could be considered a metric. In this report, we more frequently refer to a metric to imply a specific measure, which may eventually have a goal associated with it.

[10]   *Average gain* in TABE grade-equivalent scores is widely used but problematic (see Lindholm-Leary and Hargett, 2006, and Wenger et al., 2017).

eates the inputs, processes or activities, expected outputs, and desired outcomes of a specific program designed to address a problem (Shakman and Rodriguez, 2015). A logic model builds on a TOC, but it includes more information to develop metrics or indicators to monitor progress in implementation. To this end, the ChalleNGe logic model includes a detailed list of longer-term outcomes that we might expect to see in ChalleNGe graduates and that might ultimately form the basis of evaluating the program's effectiveness.[11] We included detailed information about these tools in our two earlier reports (Wenger et al., 2017; Wenger, Constant, and Cottrell, 2018), but we also include the logic model in Chapter Three of this report.

Next, we developed a detailed site-visit protocol (based partly on the logic model) and set up a schedule that allows us to visit each ChalleNGe site over the course of this project. In 2018, we completed 14 site visits. During each visit, we interviewed program leadership and staff and we collected detailed information on the program and any specific initiatives undertaken at a particular site. Among the topics we covered at each visit are the site's mission and general approach, practices used to recruit potential cadets, training of mentors, instructional practices, information about occupational training offered to cadets, placement strategies, data collection strategies, and disciplinary policies (see Table 1.1 for a full list of topics and the main source of information for each topic). It should be noted that staff also comment on other aspects of the program: for example, at one site visit, the commandant supervising the cadre staff who oversee the cadets commented on the logistical alignment of the physical training activities and the instructional program. Because our focus is on developing metrics that capture long-term outcomes, we also ask staff for their input on the types of metrics they would like to learn more about. One common request from staff is for more-detailed information on longer-term outcomes, such as the amount of postsecondary education graduates typically accrue or graduates' ability to find and retain jobs that provide sufficient resources.

The RAND research team is currently working to comprehensively analyze the data collected from sites to generate themes across the key topics listed in Table 1.1. We will provide a more complete reporting of the themes, and the findings supporting those themes, in the project's final report, to be published in summer 2020 after the completion of all site visits. Here, we present findings related to mentoring from the site visits conducted to date.

In addition to site visits and our ongoing analytic work, the study team has identified opportunities to support program sites that are conducting pilot projects. These pilots will be run primarily by the selected program site; the RAND study team will help pilot sites develop tools to assess the implementation and results of the pilot. Currently, these pilots include approaches to collecting data from graduates beyond

---

[11] For more information on logic models, see (among others) Knowlton and Phillips, 2009, and Shakman and Rodriguez, 2015.

Table 1.1
Topics Covered in the Site Visit Protocol and Main Sources of Information

Topics Covered	Main Sources of Information
Program mission	Director and deputy director
Community relationship, parent relationship	Director and deputy director
Staffing	Director and deputy director
Cadet outcomes	Director and deputy director
Finance and resources	Director and deputy director
Recruiting	Recruitment, placement, and mentoring staff
Mentoring and mentor training	Recruitment, placement, and mentoring staff
Cadet discipline	Commandant (head of the cadre staff)
Graduate placement	Recruitment, placement, and mentoring staff
Curriculum and instruction, occupational training	Instructional staff
Data management	Management information systems lead
Desired indicators	All staff

the one-year Post-Residential Phase, improving mentor training, and measuring socioemotional skills attained while enrolled in the ChalleNGe program. The RAND study team will help sites document the results of pilot implementation and will make the findings available to all sites. We describe our analytic plans in more detail in Chapter Three.

Organization of This Report

The remainder of this report consists of three chapters and an appendix:

- Chapter Two provides a snapshot of the ChalleNGe program in 2017–2018. This snapshot includes information about recent classes, comparable with information in past reports, as well as information on the proportion of cadets meeting key TABE levels, cadets' contributions to their communities, placement rates after cadets leave the program, details about a few key aspects of each program, and analyses of trends over time.
- Chapter Three discusses our initial framework for measuring the longer-term outcomes of the program. This chapter presents the program logic model and describes ongoing and planned analyses in support of measuring longer-term outcomes.

- Chapter Four presents concluding thoughts.
- The appendix includes a complete list of the ChalleNGe programs and detailed information collected from each program.

CHAPTER TWO

Data and Analyses: 2017 ChalleNGe Classes

To document progress across ChalleNGe sites, we requested and received a variety of quantitative information from each site for the cadet classes enrolled in 2017. To facilitate and ensure consistency of data collection, RAND researchers provided a spreadsheet template to each site. Sites filled in the required information and then transferred the information to RAND researchers (we requested no individually identifying information) through a secure file transfer protocol link. Sites that use one common database were able to pull the information in an automated manner. However, because not all sites use a common database, we are currently exploring options to lower the burden of future data collections.

At the time of our data collection (August–October 2018), the Georgia-Milledgeville site, the California-Discovery site, and the Tennessee ChalleNGe site, all of which were newly operational during the last data collection in 2017, had accrued enough information to be included in this report. The Puerto Rico site, which was closed during part of 2017 because of damage sustained from Hurricane Maria, also was able to provide data for this report. Since we completed our data collection, the two Texas sites have consolidated into a single site. We provide data separately from these two sites, but future reports will include information on only a single Texas site. A previous report (Wenger, Constant, and Cottrell, 2018) included data from 36 ChalleNGe sites; this report includes data from all 39 sites. As a result, our sample differs slightly from the sample used in the 2018 report.

To ensure data fidelity, we implemented several key procedures as part of our quality assurance process, including

- confirming with sites when part of their data was incomplete or missing
- exploring outliers
- comparing counts and averages across sites and classes
- comparing trends by site and classes against previous ChalleNGe reports
- comparing site data with program-wide data to ensure broad consistency.

Despite our quality assurance efforts with the data, it is important to recognize that there are likely to be errors in the data due to the nature of the data collection method; we suspect that errors are especially likely to occur in the graduate placement information. During our site visits, program administrators frequently acknowledged the difficulty of obtaining and verifying placement data from graduates and their mentors, and the further out after graduation, the more difficult it becomes. This difficulty likely limited the interpretation of the placement information in this report, as well as the use of placement information, because program sites collect this information to evaluate their success. We are currently developing a pilot program to test several ideas that have the potential to increase reporting by mentors. We will document the results of this pilot program in a future report.

We begin by presenting a summary of the information from all reporting sites. These metrics serve to measure overall progress in the ChalleNGe program in terms of the number of young people who participated in ChalleNGe programs in 2017. We also include tallies of the total number of academic credentials awarded, the hours and value of community service documented, and the overall placement rates. We then present this information in a less-aggregated manner, for each site and by class. In the next section of this chapter, we present a detailed analysis of the data on cadets' TABE scores; we use RAND-developed metrics to show the number of cadets who achieve key TABE milestones. We then present information on site-level policies and programs. We finish by presenting some time trends, using information from current and previous data collections. As we gather additional data across a longer time frame, we will continue to track trends over time.

Cross-Site Metrics for the 2017 Classes

Below we present summary information on the cadets who entered a ChalleNGe program in 2017. In total, there were 19,720 applications for programs that began during 2017. Based on site-specific enrollment criteria, 13,457 young people were accepted by, and chose to enroll in, a ChalleNGe program. Of the 13,457 cadets who enrolled, 9,759 (73 percent) *graduated* from the 5.5-month Residential Phase of ChalleNGe. Because different ChalleNGe sites begin and end their programs at different times during the year, we define *2017 participants* as those who attended a ChalleNGe program that started in 2017. In some cases, cadets may have applied in 2016 (e.g., to enter a program that began in January 2017). In most cases, cadets graduated in 2017, but a few programs spanned the 2017–2018 calendar years. Table 2.1 provides a summary of several key ChalleNGe statistics, across all sites. Tables 2.2–2.17 provide site-level information on a variety of measures.

Tables 2.2 and 2.3 provide information on the number of applicants, enrollees, credentials awarded, and graduates for each site and the two classes participating in 2017. Tables 2.4–2.12 present the TABE scores for each site and class, and Tables 2.13–2.17 present data on other core components, such as responsible citizenship, commu-

Table 2.1
ChalleNGe Statistics, 1993–2017

Challenge Statistics	1993–2016[a]	2017[b]	1993–2017
Applicants	369,741	19,720	389,461
Enrollees	208,204	13,457	221,661
Graduates	155,239	9,759	164,998
Academic credentials[c]	96,792	3,891	100,683
Service hours to communities	10,508,570	599,991	11,108,561
Value of service hours	$209,934,295	$14,374,555	$224,308,850

[a] In the 2017 report, counts excluded Puerto Rico Classes 46 and 47 because the site was not operational during the time of data collection due to Hurricane Maria in September 2017. In this table, counts have since been adjusted to include Puerto Rico Classes 46 and 47.

[b] Information in this column was reported by the sites in fall 2018 and covers Classes 48 and 49; these classes began in 2017. Applicants include all who completed an application.

[c] The academic credentials row reflects graduated cadets who received a GED, HiSET, or Test Assessing Secondary Completion (TASC) credential or a high school diploma (limited to one credential per cadet). Several sites did not report credentials; see Table 2.3 for more information. A broader definition of credentials would include high school credits awarded; by this definition, 6,212 cadets received at least one credential. Additionally, programs may have reported the total number of academic credentials for earlier classes rather than limiting credentials to one per cadet. Therefore, these numbers (and those in Table 2.3) may not be comparable with those documented in reports on ChalleNGe classes graduating prior to 2015.

nity service, and physical fitness. These tables provide a detailed sense of each site's progress on multiple metrics. Table A.1 in the appendix provides additional information on each site, including abbreviations of site names. Tables A.2–A.41 in the appendix provide more-detailed information on each ChalleNGe site, including information on staffing and funding; dates when classes began and ended; measures of physical fitness, responsible citizenship, and service to community; and detailed placement information on ChalleNGe graduates.

In some cases, individual data items are left blank in tables in this chapter and the appendix. When this occurs, we note the specific reason. Some sites failed to report specific data elements. In other cases, information was not yet available; for example, Class 49 graduates had not completed the Post-Residential Phase at the time of our data collection. Program administrators were unable to collect—and thus unable to report to the RAND team—12-month placement data on these cadets. Instead, this information will be collected during the next data collection effort.

All subsequent tables in this chapter include information for each site and class. Full names, locations, and abbreviations for the ChalleNGe sites are presented in

Table 3.1 in Chapter Three and in the appendix. TABE scores calculated in Tables 2.4–2.12 include only cadets who graduated from ChalleNGe. The tables are organized by the following metrics:

- **Numbers of applicants and graduates, including the targeted number of graduates (Table 2.2).** The targeted number of graduates is a key metric for ChalleNGe sites because it is considered in setting their budgets.
- **Credentials awarded (Table 2.3).** Table 2.3 includes a tally of credentials awarded by each site. To better determine the proportion of graduates who received at least one credential, we requested that sites report only a single credential for each graduate. Therefore, graduates who received high school credits and a high school diploma are listed as having received only a diploma; those who received a GED or HiSET certification and high school credits are listed as receiving only high school credits (this second case is quite rare). As noted, a few sites reported these data in an inconsistent manner. Table 2.1 includes only the more-restrictive definition of credentials—passing a standardized test or receiving a high school diploma. Based on this definition, about 40 percent of ChalleNGe graduates received a credential. But if we also include high school credits as credentials, over 60 percent of graduates received at least one credential. Table 2.3 includes this broader set of credentials.
- **TABE scores (Tables 2.4–2.12).** We collected information on the total TABE battery but also on three specific subtests: math, language, and reading. We report information on each subtest. We also report additional information on TABE scores for all cadets in a later subsection of this chapter. TABE scores are reported as grade equivalents; for example, a score of 7.5 indicates that the test taker performed similarly to a typical student at the fifth month of seventh grade. Cadets generally achieved higher TABE scores at the end of ChalleNGe than at the beginning, across sites.
- **Responsible citizenship (Tables 2.13 and 2.14).** Metrics of responsible citizenship include registration for voting (all cadets) and registration for the Selective Service (male cadets). The majority of sites registered 100 percent of eligible cadets for voting and Selective Service.
- **Community service (Table 2.15).** We report the average hours of community service per cadet, as well as the value of that service. The value of community service is calculated using published figures at the state level for 2015, which are available online (Independent Sector, 2016). The value of community service was calculated in the same manner in the past three annual reports (Wenger, Constant, and Cottrell, 2018; Wenger et al., 2017; National Guard Youth ChalleNGe, 2015). Each cadet contributed 40 to 130 hours of community service. Cadets from Virginia, Hawaii, and North Carolina-Salemburg contributed the highest number of hours.

- **Physical fitness (Tables 2.16 and 2.17).** We report one-mile run times and push-ups for Classes 48 and 49. Cadets were able to perform more than 15 additional push-ups and ran about two minutes faster at the end of ChalleNGe. We include additional tabulations on these metrics in a later subsection of this chapter. In the 2018 report (Wenger, Constant, and Cottrell, 2018), we included more-detailed data on changes in cadets' BMI and on the proportion of cadets achieving various levels of fitness. In the interest of brevity, we do not include similar information in this report. However, we did collect and analyze the information, and we found results on these health- and fitness-related outcomes that were very similar to 2018's results.

Table 2.2
Applicants and Graduates (Classes 48 and 49)

Site	Residential Class 48			Residential Class 49		
	Target	Applied	Graduates	Target	Applied	Graduates
All Sites		9,650	4,844		10,070	4,915
AK	185	394	180	150	331	146
AR	100	250	103	100	261	88
CA-DC	100	214	105	125	257	126
CA-LA	180	326	172	190	316	193
CA-SL	185	294	186	190	334	195
D.C.	100	125	29	85	101	45
FL	150	203	163	150	211	158
GA-FG	*	277	202	*	255	135
GA-FS	212	298	222	213	290	224
GA-MV	150	168	76	150	205	107
HI-BP	125	150	97	125	185	105
HI-HI	75	79	62	75	105	67
ID	100	165	104	105	189	115
IL	200	278	124	175	364	138
IN	100	104	104	100	97	97
KY-FK	*	147	96	*	136	72
KY-HN	100	149	77	100	143	82
LA-CB	250	428	202	250	478	250
LA-CM	200	333	207	200	341	213

Table 2.2—Continued

Site	Residential Class 48			Residential Class 49		
	Target	Applied	Graduates	Target	Applied	Graduates
LA-GL	250	464	254	250	436	214
MD	100	220	100	100	253	102
MI	114	195	116	114	248	119
MS	200	444	169	200	577	200
MT	100	137	80	100	127	80
NC-NL	100	246	95	100	366	99
NC-S	400	261	88	400	289	105
NJ	100	286	82	100	254	60
NM	100	156	93	100	168	107
OK	110	374	101	110	451	114
OR	125	293	131	125	198	123
PR	200	280	220	200	247	216
SC	100	186	112	100	175	103
TN	N/A	N/A	N/A	*	51	23
TX-E	100	219	79	100	224	55
TX-W	88	247	41	87	145	61
VA	101	240	100	116	221	116
WA	135	243	143	135	296	141
WI	100	253	92	100	275	108
WV	150	364	151	150	348	153
WY	60	160	86	60	122	60

NOTE: Information in this table was reported by the sites in fall 2018 and covers Classes 48 and 49, which began in 2017. Target columns represent the program's graduation goal. Additional information is available in the appendix.

* Did not report.

N/A = not available.

Table 2.3
Number of ChalleNGe Graduates and Number of Graduates by Type of Credential Awarded, by Site (Classes 48 and 49)

Site	Residential Class 48				Residential Class 49			
	Number of Graduates from ChalleNGe	Number Receiving GED, HiSET, or TASC	Number Receiving High School Credits	Number Receiving High School Diploma	Number of Graduates from ChalleNGe	Number Receiving GED, HiSET, or TASC	Number Receiving High School Credits	Number Receiving High School Diploma
AK	180	102	0	6	146	68	0	3
AR	103	34	0	0	88	35	0	0
CA-DC	105	0	75	30	126	0	96	30
CA-LA	172	0	155	17	193	0	175	18
CA-SL	186	0	149	37	195	0	149	46
D.C.	29	0	0	6	45	0	0	12
FL	163	111	22	4	158	96	20	4
GA-FG	202	*	*	*	135	*	*	*
GA-FS	222	143	0	23	224	143	0	11
GA-MV	76	32	10	1	107	44	11	3
HI-BP	97	97	0	0	105	104	0	0
HI-HI	62	0	0	62	67	0	0	67
ID	104	0	93	11	115	0	99	16
IL	124	81	0	0	138	75	0	0
IN	104	63	0	0	97	53	0	0
KY-FK	96	*	*	*	72	*	*	*
KY-HN	77	0	75	0	82	0	78	0
LA-CB	202	63	0	0	250	84	0	0
LA-CM	207	77	0	0	213	65	0	0
LA-GL	254	92	0	0	214	87	5	0
MD	100	0	0	45	102	0	0	56
MI	116	0	107	9	119	0	94	25
MS	169	0	0	103	200	0	0	108
MT	80	31	0	0	80	29	0	0
NC-NL	95	33	3	0	99	24	0	24

Table 2.3—Continued

Site	Residential Class 48				Residential Class 49			
	Number of Graduates from ChalleNGe	Number Receiving GED, HiSET, or TASC	Number Receiving High School Credits	Number Receiving High School Diploma	Number of Graduates from ChalleNGe	Number Receiving GED, HiSET, or TASC	Number Receiving High School Credits	Number Receiving High School Diploma
NC-S	88	54	0	0	105	*	*	*
NJ	82	0	0	29	60	0	0	24
NM	93	61	0	0	107	63	0	0
OK	101	0	68	31	114	6	83	24
OR	131	0	118	13	123	0	107	16
PR	220	*	*	*	216	0	1	215
SC	112	53	0	0	103	43	0	0
TN	N/A	N/A	N/A	N/A	23	8	0	0
TX-E	79	0	64	15	55	2	40	12
TX-W	41	0	31	10	61	0	52	9
VA	100	39	0	0	116	41	0	0
WA	143	0	139	0	141	0	137	0
WI	92	40	0	28	108	43	0	38
WV	151	0	20	129	153	0	27	126
WY	86	42	15	2	60	32	3	0

NOTE: Information in this table was reported by the sites in fall 2018 and covers Classes 48 and 49, which began in 2017. Credentials awarded include those awarded during the course of the ChalleNGe Residential Phase. Counts reflect a single credential per cadet. Cadets with multiple credentials are assigned based on the hierarchy of high school diploma, then high school credits, then GED/HiSet/TASC. At the Idaho ChalleNGe program, those who received GEDs also received high school credits, although the credits were not used. In New Jersey, ChalleNGe graduates who pass the GED are awarded a state high school diploma. In West Virginia, ChalleNGe graduates who pass the state standardized test are awarded a state high school diploma. The Wisconsin program generates a pathway for all credentialing options awarded through the Wisconsin Department of Instruction and associated school districts, including credit recovery, GED, a high school equivalency diploma, and a high school diploma. Additional information on each ChalleNGe site is available in the appendix.

* Did not report.

N/A = not available.

Table 2.4
Average TABE Math Score at Beginning and End of Residential Phase and Gain of ChalleNGe Graduates, by Site (Classes 48 and 49)

Site	Residential Class 48			Residential Class 49		
	Pre-TABE	Post-TABE	Gain (+/−)	Pre-TABE	Post-TABE	Gain (+/−)
All Sites	6.5	8.7	2.2	6.3	8.6	2.3
AK	7.6	9.3	1.7	7.4	8.9	1.5
AR	7.6	7.8	0.2	6.2	8.3	2.1
CA-DC	6.2	8.1	1.9	6.4	8.4	2.0
CA-LA	4.7	8.0	3.3	4.8	8.0	3.2
CA-SL	6.9	8.2	1.3	6.6	8.1	1.5
D.C.	5.6	7.2	1.6	5.1	7.3	2.2
FL	7.0	9.2	2.2	7.1	9.7	2.6
GA-FG	5.5	7.6	2.1	6.0	7.7	1.7
GA-FS	6.7	10.4	3.7	7.0	10.3	3.3
GA-MV	5.9	7.4	1.5	6.1	8.1	2.0
HI-BP	5.5	7.6	2.1	6.2	8.0	1.8
HI-HI	5.5	5.9	0.4	5.2	5.7	0.5
ID	7.5	8.9	1.4	7.2	9.5	2.3
IL	5.7	9.9	4.2	5.3	9.2	3.9
IN	7.8	9.4	1.6	7.3	8.7	1.4
KY-FK	6.1	8.3	2.2	5.1	7.8	2.7
KY-HN	2.7	7.5	4.8	4.7	8.4	3.7
LA-CB	5.6	9.6	4.0	5.9	9.4	3.5
LA-CM	6.3	9.5	3.2	6.2	9.0	2.8
LA-GL	6.8	9.2	2.4	7.0	9.2	2.2
MD	5.7	8.6	2.9	5.3	8.5	3.2
MI	6.7	7.3	0.6	6.6	7.0	0.4
MS	6.6	10.0	3.4	6.1	9.4	3.3
MT	7.8	9.1	1.3	7.7	8.8	1.1
NC-NL	6.4	7.2	0.8	6.7	7.9	1.2
NC-S	6.3	8.1	1.8	6.1	8.2	2.1

Table 2.4—Continued

Site	Residential Class 48			Residential Class 49		
	Pre-TABE	Post-TABE	Gain (+/−)	Pre-TABE	Post-TABE	Gain (+/−)
NJ	6.4	8.8	2.4	6.4	10.2	3.8
NM	6.2	8.8	2.6	6.0	8.4	2.4
OK	7.5	9.0	1.5	8.1	9.5	1.4
OR	7.1	9.0	1.9	7.2	8.8	1.6
PR	*	*	*	4.1	6.9	2.8
SC	6.4	8.6	2.2	7.8	8.1	0.3
TN	N/A	N/A	N/A	4.3	6.3	2.0
TX-E	*	*	*	6.7	8.5	1.8
TX-W	7.5	7.0	−0.5	7.3	8.3	1.0
VA	6.2	5.5	−0.7	5.9	7.0	1.1
WA	6.1	9.2	3.1	5.9	8.8	2.9
WI	8.6	9.1	0.5	8.1	9.0	0.9
WV	5.4	7.8	2.4	6.1	8.0	1.9
WY	8.1	9.4	1.3	6.8	8.3	1.5

NOTE: Information in this table was reported by the sites in fall 2018 and covers Classes 48 and 49, which began in 2017.
* Did not report.
N/A = not available.

Table 2.5
Average TABE Reading Score at Beginning and End of Residential Phase and Gain of ChalleNGe Graduates, by Site (Classes 48 and 49)

Site	Residential Class 48			Residential Class 49		
	Pre-TABE	Post-TABE	Gain (+/−)	Pre-TABE	Post-TABE	Gain (+/−)
All Sites	7.5	9.0	1.5	7.5	9.1	1.6
AK	8.1	9.3	1.2	7.1	8.8	1.7
AR	8.5	8.9	0.4	8.8	9.4	0.6
CA-DC	7.9	9.3	1.4	6.7	9.5	2.8
CA-LA	5.5	9.2	3.7	5.6	9.4	3.8
CA-SL	7.8	9.0	1.2	7.5	9.0	1.5
D.C.	6.5	7.6	1.1	7.4	8.2	0.8

Table 2.5—Continued

Site	Residential Class 48			Residential Class 49		
	Pre-TABE	Post-TABE	Gain (+/−)	Pre-TABE	Post-TABE	Gain (+/−)
FL	7.7	10.9	3.2	8.1	10.0	1.9
GA-FG	6.2	8.0	1.8	6.7	8.2	1.5
GA-FS	7.5	10.6	3.1	7.6	9.8	2.2
GA-MV	7.1	6.7	−0.4	7.0	8.1	1.1
HI-BP	6.4	8.0	1.6	8.1	8.8	0.7
HI-HI	6.4	5.0	−1.4	6.8	5.9	−0.9
ID	8.5	9.7	1.2	8.5	10.0	1.5
IL	8.5	8.8	0.3	8.1	8.6	0.5
IN	8.0	10.0	2.0	7.8	8.8	1.0
KY-FK	6.3	7.1	0.8	5.2	8.2	3.0
KY-HN	3.1	6.1	3.0	5.8	7.3	1.5
LA-CB	6.4	9.8	3.4	7.7	9.6	1.9
LA-CM	7.8	9.6	1.8	7.8	9.3	1.5
LA-GL	8.4	9.4	1.0	8.6	9.7	1.1
MD	6.6	9.2	2.6	6.6	9.5	2.9
MI	6.6	6.0	−0.6	6.6	6.1	−0.5
MS	8.1	10.6	2.5	7.8	10.0	2.2
MT	8.1	9.0	0.9	8.4	8.8	0.4
NC-NL	7.3	7.9	0.6	7.5	8.7	1.2
NC-S	7.6	9.2	1.6	7.5	9.3	1.8
NJ	7.7	8.9	1.2	6.9	9.6	2.7
NM	7.2	9.3	2.1	8.1	9.8	1.7
OK	7.6	8.8	1.2	7.9	8.7	0.8
OR	8.8	10.0	1.2	8.9	9.6	0.7
PR	*	*	*	5.1	10.0	4.9
SC	8.0	8.8	0.8	7.8	7.8	0.0
TN	N/A	N/A	N/A	7.6	8.8	1.2
TX-E	*	*	*	7.8	9.2	1.4
TX-W	8.8	8.1	−0.7	7.5	7.3	−0.2

Table 2.5—Continued

Site	Residential Class 48			Residential Class 49		
	Pre-TABE	Post-TABE	Gain (+/−)	Pre-TABE	Post-TABE	Gain (+/−)
VA	6.9	6.7	−0.2	6.7	7.2	0.5
WA	7.3	9.4	2.1	7.7	9.5	1.8
WI	8.6	9.3	0.7	8.9	8.8	−0.1
WV	7.6	8.6	1.0	7.9	9.6	1.7
WY	9.0	9.9	0.9	8.9	9.3	0.4

NOTE: Information in this table was reported by the sites in fall 2018 and covers Classes 48 and 49, which began in 2017.

* Did not report.

N/A = not available.

Table 2.6
Average TABE Total Battery Score at Beginning and End of Residential Program and Gain of ChalleNGe Graduates, by Site (Classes 48 and 49)

Site	Residential Class 48			Residential Class 49		
	Pre-TABE	Post-TABE	Gain (+/−)	Pre-TABE	Post-TABE	Gain (+/−)
All Sites	6.7	8.9	2.2	6.7	8.8	2.1
AK	7.7	9.7	2.0	7.3	9.0	1.7
AR	8.3	8.7	0.4	8.0	9.3	1.3
CA-DC	*	*	*	*	*	*
CA-LA	5.4	9.3	3.9	5.3	9.4	4.1
CA-SL	7.2	8.8	1.6	6.8	8.8	2.0
D.C.	5.2	7.3	2.1	5.5	7.3	1.8
FL	7.1	10.6	3.5	7.4	10.0	2.6
GA-FG	5.4	7.6	2.2	5.6	7.7	2.1
GA-FS	6.7	10.4	3.7	7.0	10.0	3.0
GA-MV	6.4	6.7	0.3	6.4	7.9	1.5
HI-BP	5.4	7.4	2.0	6.6	8.0	1.4
HI-HI	5.3	5.0	−0.3	5.6	5.5	−0.1
ID	7.8	9.6	1.8	7.8	9.8	2.0
IL	7.5	9.0	1.5	7.0	8.3	1.3
IN	7.6	9.7	2.1	7.5	8.8	1.3

Table 2.6—Continued

Site	Residential Class 48			Residential Class 49		
	Pre-TABE	Post-TABE	Gain (+/−)	Pre-TABE	Post-TABE	Gain (+/−)
KY-FK	5.6	7.2	1.6	4.3	7.3	3.0
KY-HN	2.3	6.0	3.7	4.4	7.4	3.0
LA-CB	5.7	9.8	4.1	6.6	9.5	2.9
LA-CM	6.8	9.6	2.8	6.8	9.2	2.4
LA-GL	7.6	9.5	1.9	8.0	9.8	1.8
MD	5.6	8.7	3.1	5.6	8.9	3.3
MI	6.1	6.3	0.2	6.3	6.4	0.1
MS	7.3	10.7	3.4	7.0	10.1	3.1
MT	7.7	8.9	1.2	7.8	8.8	1.0
NC-NL	6.0	7.2	1.2	6.7	8.3	1.6
NC-S	6.7	8.4	1.7	6.4	8.6	2.2
NJ	6.8	8.9	2.1	6.8	10.0	3.2
NM	6.5	8.9	2.4	6.8	9.0	2.2
OK	7.1	8.6	1.5	7.6	8.8	1.2
OR	7.7	9.6	1.9	7.8	9.1	1.3
PR	*	*	*	4.5	8.4	3.9
SC	7.4	8.9	1.5	8.0	7.8	−0.2
TN	N/A	N/A	N/A	*	*	*
TX-E	*	*	*	*	8.5	*
TX-W	7.9	7.2	−0.7	6.9	7.3	0.4
VA	6.2	6.4	0.2	6.3	6.9	0.6
WA	6.4	9.2	2.8	6.3	9.2	2.9
WI	8.2	9.3	1.1	8.1	8.8	0.7
WV	6.1	8.1	2.0	6.6	8.5	1.9
WY	8.5	9.9	1.4	7.7	8.9	1.2

NOTE: Information in this table was reported by the sites in fall 2018 and covers Classes 48 and 49, which began in 2017.

* Did not report.

N/A = not available.

Table 2.7
Distribution of TABE Math Scores of ChalleNGe Graduates at Beginning and End of Residential Phase, by Site (Class 48)

Site	Pre-TABE			Post-TABE		
	Elementary (Grades 1–6)	Middle School (Grades 7–8)	High School (Grades 9–12)	Elementary (Grades 1–6)	Middle School (Grades 7–8)	High School (Grades 9–12)
All Sites	2,314	1,234	911	994	1,403	1,996
AK	78	38	64	30	50	100
AR	35	34	34	33	34	34
CA-DC	62	21	22	35	29	41
CA-LA	123	32	17	60	50	62
CA-SL	95	50	41	51	67	68
D.C.	16	11	2	11	13	5
FL	69	48	46	14	63	86
GA-FG	93	34	18	29	58	33
GA-FS	96	88	38	6	53	162
GA-MV	37	26	7	21	26	12
HI-BP	64	18	14	40	29	27
HI-HI	44	8	10	42	12	8
ID	38	36	30	20	38	46
IL	82	21	21	6	43	75
IN	35	34	35	21	20	62
KY-FK	60	19	15	27	32	35
KY-HN	66	2	0	19	17	15
LA-CB	127	52	23	27	55	120
LA-CM	108	60	38	36	49	121
LA-GL	113	94	47	31	87	135
MD	69	17	14	22	32	44
MI	55	36	25	51	31	34
MS	92	41	36	9	51	109
MT	28	27	25	14	19	44
NC-NL	48	29	15	42	28	23
NC-S	47	27	14	25	29	33

Table 2.7—Continued

Site	Pre-TABE			Post-TABE		
	Elementary (Grades 1–6)	Middle School (Grades 7–8)	High School (Grades 9–12)	Elementary (Grades 1–6)	Middle School (Grades 7–8)	High School (Grades 9–12)
NJ	36	28	18	17	29	36
NM	46	38	9	8	45	40
OK	37	31	33	18	31	52
OR	58	39	34	22	47	62
PR	*	*	*	*	*	*
SC	60	25	22	26	31	49
TN	N/A	N/A	N/A	N/A	N/A	N/A
TX-E	*	*	*	*	*	*
TX-W	13	14	14	17	15	9
VA	59	21	20	64	34	2
WA	80	35	26	19	48	72
WI	22	29	41	14	35	43
WV	99	40	12	54	49	48
WY	24	31	31	13	24	49

NOTE: Information in this table was reported by the sites in fall 2018 and covers Class 48.

* Did not report.

N/A = not available.

Table 2.8
Distribution of TABE Math Scores of ChalleNGe Graduates at Beginning and End of Residential Phase, by Site (Class 49)

Site	Pre-TABE			Post-TABE		
	Elementary (Grades 1–6)	Middle School (Grades 7–8)	High School (Grades 9–12)	Elementary (Grades 1–6)	Middle School (Grades 7–8)	High School (Grades 9–12)
All Sites	2,553	1,284	919	1,071	1,576	2,072
AK	66	35	45	28	48	70
AR	50	15	22	9	42	23
CA-DC	70	28	28	28	46	47
CA-LA	140	37	13	56	61	76
CA-SL	101	57	37	57	65	73

Table 2.8—Continued

Site	Pre-TABE			Post-TABE		
	Elementary (Grades 1–6)	Middle School (Grades 7–8)	High School (Grades 9–12)	Elementary (Grades 1–6)	Middle School (Grades 7–8)	High School (Grades 9–12)
D.C.	34	10	1	16	18	11
FL	58	58	42	7	50	101
GA-FG	45	27	12	18	27	22
GA-FS	83	96	43	8	56	158
GA-MV	57	29	18	29	37	41
HI-BP	58	33	14	34	38	33
HI-HI	48	12	7	40	14	10
ID	55	25	35	16	31	68
IL	95	20	23	16	58	64
IN	39	28	30	25	26	45
KY-FK	42	16	4	19	24	20
KY-HN	60	13	9	24	23	35
LA-CB	141	70	39	46	58	146
LA-CM	107	69	37	36	75	102
LA-GL	87	72	54	37	53	121
MD	72	19	8	21	37	39
MI	55	34	27	48	37	33
MS	110	62	28	13	75	112
MT	33	17	30	17	24	36
NC-NL	51	22	26	39	22	38
NC-S	26	17	6	13	14	25
NJ	28	20	10	4	14	41
NM	56	37	14	23	47	37
OK	33	38	43	15	39	60
OR	58	29	36	26	45	52
PR	194	18	3	84	100	30
SC	25	39	27	20	43	32
TN	20	2	0	13	6	3

Table 2.8—Continued

Site	Pre-TABE			Post-TABE		
	Elementary (Grades 1–6)	Middle School (Grades 7–8)	High School (Grades 9–12)	Elementary (Grades 1–6)	Middle School (Grades 7–8)	High School (Grades 9–12)
TX-E	25	13	13	12	17	25
TX-W	28	13	19	17	17	25
VA	74	20	22	47	37	25
WA	79	38	20	25	48	64
WI	28	40	40	19	36	53
WV	96	35	22	51	49	53
WY	26	21	12	15	19	23

NOTE: Information in this table was reported by the sites in fall 2018 and covers Class 49.

Table 2.9
Distribution of TABE Reading Scores of ChalleNGe Graduates at Beginning and End of Residential Phase, by Site (Class 48)

Site	Pre-TABE			Post-TABE		
	Elementary (Grades 1–6)	Middle School (Grades 7–8)	High School (Grades 9–12)	Elementary (Grades 1–6)	Middle School (Grades 7–8)	High School (Grades 9–12)
All Sites	1,668	1,222	1557	805	1,184	2,419
AK	58	35	87	22	51	107
AR	23	33	47	17	31	53
CA-DC	42	21	42	22	17	66
CA-LA	115	32	25	30	35	107
CA-SL	54	64	68	24	64	98
D.C.	17	7	5	9	9	11
FL	60	34	69	6	20	137
GA-FG	80	34	30	30	42	48
GA-FS	79	71	72	15	33	174
GA-MV	32	17	21	35	18	21
HI-BP	47	31	18	27	30	39
HI-HI	30	17	13	44	9	9
ID	26	28	50	13	27	64
IL	28	41	55	21	40	63

Table 2.9—Continued

Site	Pre-TABE			Post-TABE		
	Elementary (Grades 1–6)	Middle School (Grades 7–8)	High School (Grades 9–12)	Elementary (Grades 1–6)	Middle School (Grades 7–8)	High School (Grades 9–12)
IN	29	38	37	12	23	68
KY-FK	49	21	24	34	33	28
KY-HN	60	6	2	27	15	8
LA-CB	97	63	42	27	46	129
LA-CM	65	60	80	27	49	130
LA-GL	66	70	118	39	62	152
MD	48	28	23	18	27	52
MI	56	24	36	71	21	24
MS	48	46	75	12	28	129
MT	26	20	33	9	25	43
NC-NL	33	29	26	27	29	36
NC-S	35	21	31	13	28	47
NJ	31	19	32	16	25	41
NM	33	35	25	10	30	53
OK	40	20	41	23	27	51
OR	26	41	64	6	39	86
PR	*	*	*	*	*	*
SC	32	34	41	19	36	51
TN	N/A	N/A	N/A	N/A	N/A	N/A
TX-E	*	*	*	*	*	*
TX-W	9	13	19	10	14	17
VA	43	34	22	21	73	6
WA	58	37	46	16	39	84
WI	22	26	44	13	27	52
WV	53	51	47	33	41	77
WY	18	21	47	7	21	58

NOTE: Information in this table was reported by the sites in fall 2018 and covers Class 48.

* Did not report.

N/A = not available.

Table 2.10
Distribution of TABE Reading Scores of ChalleNGe Graduates at Beginning and End of Residential Phase, by Site (Class 49)

Site	Pre-TABE			Post-TABE		
	Elementary (Grades 1–6)	Middle School (Grades 7–8)	High School (Grades 9–12)	Elementary (Grades 1–6)	Middle School (Grades 7–8)	High School (Grades 9–12)
All Sites	1,752	1,288	1,691	838	1,289	2,586
AK	58	46	42	29	45	72
AR	14	25	45	9	19	46
CA-DC	50	33	42	16	28	82
CA-LA	120	40	32	17	58	118
CA-SL	68	63	64	28	62	105
D.C.	15	12	18	11	12	22
FL	48	41	69	10	44	104
GA-FG	40	21	23	17	22	28
GA-FS	74	75	73	26	44	152
GA-MV	50	23	34	33	28	46
HI-BP	26	37	41	18	36	51
HI-HI	32	18	17	40	15	12
ID	35	26	54	10	30	75
IL	34	49	55	24	52	62
IN	26	35	36	20	24	52
KY-FK	39	10	15	15	22	26
KY-HN	49	13	20	33	21	28
LA-CB	85	67	96	45	52	153
LA-CM	80	44	88	31	57	125
LA-GL	52	49	112	29	53	129
MD	53	22	24	13	29	55
MI	59	32	25	65	28	25
MS	71	50	79	21	45	134
MT	24	19	37	14	27	38
NC-NL	28	41	30	19	34	46
NC-S	17	18	13	4	17	31

Table 2.10—Continued

Site	Pre-TABE			Post-TABE		
	Elementary (Grades 1–6)	Middle School (Grades 7–8)	High School (Grades 9–12)	Elementary (Grades 1–6)	Middle School (Grades 7–8)	High School (Grades 9–12)
NJ	24	16	19	5	21	33
NM	25	37	45	9	28	70
OK	37	32	45	19	37	58
OR	19	42	62	11	37	75
PR	149	21	28	47	25	132
SC	21	36	27	21	40	28
TN	5	11	6	2	4	16
TX-E	16	16	20	11	13	30
TX-W	22	15	23	24	13	22
VA	53	25	38	40	34	35
WA	52	36	49	15	32	90
WI	21	33	54	18	35	55
WV	50	45	58	10	54	89
WY	11	14	33	9	12	36

NOTE: Information in this table was reported by the sites in fall 2018 and covers Class 49.

Table 2.11
Distribution of TABE Total Battery Scores of ChalleNGe Graduates at Beginning and End of Residential Phase, by Site (Class 48)

Site	Pre-TABE			Post-TABE		
	Elementary (Grades 1–6)	Middle School (Grades 7–8)	High School (Grades 9–12)	Elementary (Grades 1–6)	Middle School (Grades 7–8)	High School (Grades 9–12)
All Sites	2,083	1,186	1,086	867	1,220	2,222
AK	62	48	70	27	41	112
AR	27	29	47	24	28	49
CA-DC	*	*	*	*	*	*
CA-LA	120	33	18	28	52	92
CA-SL	72	66	48	32	64	90
D.C.	20	6	3	8	12	9
FL	66	55	42	5	27	131

Table 2.11—Continued

Site	Pre-TABE			Post-TABE		
	Elementary (Grades 1–6)	Middle School (Grades 7–8)	High School (Grades 9–12)	Elementary (Grades 1–6)	Middle School (Grades 7–8)	High School (Grades 9–12)
GA-FG	102	24	19	32	53	35
GA-FS	99	76	47	11	51	160
GA-MV	35	19	16	33	26	15
HI-BP	67	16	13	36	33	27
HI-HI	44	10	8	44	15	3
ID	37	28	39	12	31	61
IL	48	33	43	20	34	70
IN	36	33	35	12	26	65
KY-FK	64	15	17	39	27	29
KY-HN	66	2	0	31	10	10
LA-CB	123	51	28	28	42	132
LA-CM	89	65	52	29	52	125
LA-GL	88	80	86	33	65	155
MD	72	13	14	18	35	45
MI	68	22	26	70	20	26
MS	59	63	47	6	31	132
MT	30	20	30	12	26	41
NC-NL	54	22	17	38	29	27
NC-S	40	29	19	20	32	36
NJ	38	24	20	21	13	48
NM	47	32	14	12	35	46
OK	43	26	32	25	25	51
OR	48	38	45	10	41	80
PR	*	*	*	*	*	*
SC	44	32	31	18	38	50
TN	N/A	N/A	N/A	N/A	N/A	N/A
TX-E	*	*	*	*	*	*
TX-W	14	12	15	15	14	12

Table 2.11—Continued

Site	Pre-TABE			Post-TABE		
	Elementary (Grades 1–6)	**Middle School (Grades 7–8)**	**High School (Grades 9–12)**	**Elementary (Grades 1–6)**	**Middle School (Grades 7–8)**	**High School (Grades 9–12)**
VA	59	22	19	41	45	14
WA	74	38	29	15	48	76
WI	24	33	35	10	33	49
WV	76	56	19	43	46	62
WY	28	15	43	9	20	57

NOTE: Information in this table was reported by the sites in fall 2018 and covers Class 48.

* Did not report.

N/A = not available.

Table 2.12
Distribution of TABE Total Battery Scores of ChalleNGe Graduates at Beginning and End of Residential Phase, by Site (Class 49)

Site	Pre-TABE			Post-TABE		
	Elementary (Grades 1–6)	**Middle School (Grades 7–8)**	**High School (Grades 9–12)**	**Elementary (Grades 1–6)**	**Middle School (Grades 7–8)**	**High School (Grades 9–12)**
All Sites	2,173	1,231	1,126	953	1,319	2,280
AK	63	34	49	28	39	78
AR	24	32	31	8	24	43
CA-DC	*	*	*	*	*	*
CA-LA	129	40	20	25	52	116
CA-SL	91	62	42	36	65	94
D.C.	33	7	5	18	16	11
FL	51	65	42	7	50	101
GA-FG	58	16	10	22	23	22
GA-FS	87	81	54	14	58	150
GA-MV	56	22	24	35	30	42
HI-BP	39	47	19	30	31	44
HI-HI	48	8	11	43	14	10
ID	47	26	42	13	25	77
IL	66	30	41	32	52	54

Table 2.12—Continued

Site	Pre-TABE			Post-TABE		
	Elementary (Grades 1–6)	Middle School (Grades 7–8)	High School (Grades 9–12)	Elementary (Grades 1–6)	Middle School (Grades 7–8)	High School (Grades 9–12)
IN	33	30	34	22	27	48
KY-FK	48	11	5	25	15	23
KY-HN	63	10	9	40	11	31
LA-CB	124	65	61	43	58	149
LA-CM	101	57	55	37	62	114
LA-GL	64	63	86	27	50	134
MD	71	15	13	19	34	44
MI	65	26	24	61	27	30
MS	84	65	51	13	53	134
MT	28	21	30	12	31	36
NC-NL	50	25	24	31	23	45
NC-S	24	16	9	12	15	25
NJ	24	20	14	4	19	36
NM	39	45	23	17	35	55
OK	47	24	43	27	32	55
OR	46	29	48	17	44	62
PR	161	24	13	41	47	95
SC	18	42	26	22	44	30
TN	*	*	*	*	*	*
TX-E	*	*	*	13	19	22
TX-W	27	16	17	23	18	18
VA	67	17	32	45	36	28
WA	71	42	24	19	43	75
WI	30	36	41	24	33	51
WV	76	43	34	35	52	66
WY	20	19	20	13	12	32

NOTE: Information in this table was reported by the sites in fall 2018 and covers Class 49.

* Did not report.

Table 2.13
Core Component Completion Among ChalleNGe Graduates: Responsible Citizenship, by Site (Class 48)

Site	Eligible to Vote	Registered to Vote	% Eligible Who Registered	Eligible for Selective Service	Registered for Selective Service	% Eligible Who Registered
All Sites	1,066	921	86%	1,292	1,189	92%
AK	27	27	100%	21	21	100%
AR	20	17	85%	41	0	0%
CA-DC	12	12	100%	12	12	100%
CA-LA	25	25	100%	36	36	100%
CA-SL	30	30	100%	19	19	100%
D.C.	6	2	33%	26	5	19%
FL	36	36	100%	31	31	100%
GA-FG	50	50	100%	58	58	100%
GA-FS	55	55	100%	44	44	100%
GA-MV	20	20	100%	15	15	100%
HI-BP	18	16	89%	39	39	100%
HI-HI	12	12	100%	29	29	100%
ID	20	20	100%	30	30	100%
IL	27	0	0%	23	23	100%
IN	15	15	100%	40	40	100%
KY-FK	13	13	100%	13	13	100%
KY-HN	13	13	100%	30	30	100%
LA-CB	34	34	100%	90	90	100%
LA-CM	37	0	0%	73	73	100%
LA-GL	81	35	43%	34	34	100%
MD	26	26	100%	41	41	100%
MI	25	25	100%	32	32	100%
MS	45	45	100%	60	60	100%
MT	11	11	100%	36	21	58%
NC-NL	11	11	100%	14	14	100%
NC-S	46	46	100%	10	10	100%
NJ	24	24	100%	19	19	100%

Table 2.13—Continued

Site	Eligible to Vote	Registered to Vote	% Eligible Who Registered	Eligible for Selective Service	Registered for Selective Service	% Eligible Who Registered
NM	23	23	100%	39	39	100%
OK	21	21	100%	35	35	100%
OR	66	66	100%	31	31	100%
PR	37	36	97%	35	35	100%
SC	25	9	36%	18	7	39%
TN	N/A	N/A	N/A	N/A	N/A	N/A
TX-E	19	17	89%	11	11	100%
TX-W	8	8	100%	8	8	100%
VA	16	10	63%	37	22	59%
WA	36	36	100%	68	68	100%
WI	28	27	96%	50	50	100%
WV	32	32	100%	31	31	100%
WY	16	16	100%	13	13	100%

NOTE: Information in this table was reported by the sites in fall 2018 and covers Class 48.

N/A = not available.

Table 2.14
Core Component Completion Among ChalleNGe Graduates: Responsible Citizenship, by Site (Class 49)

Site	Eligible to Vote	Registered to Vote	% Eligible Who Registered	Eligible for Selective Service	Registered for Selective Service	% Eligible Who Registered
All Sites	1,127	1,024	91%	1,425	1,352	95%
AK	34	34	100%	22	22	100%
AR	17	0	0%	35	19	54%
CA-DC	7	7	100%	7	7	100%
CA-LA	41	41	100%	42	42	100%
CA-SL	27	27	100%	18	18	100%
D.C.	26	26	100%	33	11	33%
FL	30	30	100%	23	23	100%
GA-FG	34	34	100%	69	69	100%
GA-FS	57	57	100%	46	46	100%

Table 2.14—Continued

Site	Eligible to Vote	Registered to Vote	% Eligible Who Registered	Eligible for Selective Service	Registered for Selective Service	% Eligible Who Registered
GA-MV	24	24	100%	22	22	100%
HI-BP	20	20	100%	59	59	100%
HI-HI	12	12	100%	31	31	100%
ID	22	22	100%	34	34	100%
IL	29	0	0%	23	23	100%
IN	14	14	100%	33	33	100%
KY-FK	14	14	100%	14	14	100%
KY-HN	9	9	100%	30	30	100%
LA-CB	33	33	100%	111	111	100%
LA-CM	41	0	0%	66	65	98%
LA-GL	43	43	100%	27	27	100%
MD	22	22	100%	34	34	100%
MI	20	20	100%	23	23	100%
MS	62	62	100%	88	88	100%
MT	14	14	100%	42	19	45%
NC-NL	21	21	100%	18	18	100%
NC-S	61	61	100%	25	25	100%
NJ	13	13	100%	8	8	100%
NM	22	22	100%	46	46	100%
OK	16	16	100%	27	27	100%
OR	76	76	100%	39	39	100%
PR	81	80	99%	69	69	100%
SC	17	9	53%	16	9	56%
TN	3	0	0%	3	0	0%
TX-E	23	20	87%	15	15	100%
TX-W	19	19	100%	20	20	100%
VA	23	22	96%	47	46	98%
WA	34	34	100%	70	70	100%
WI	27	27	100%	54	54	100%

Table 2.14—Continued

Site	Eligible to Vote	Registered to Vote	% Eligible Who Registered	Eligible for Selective Service	Registered for Selective Service	% Eligible Who Registered
WV	31	31	100%	29	29	100%
WY	8	8	100%	7	7	100%

NOTE: Information in this table was reported by the sites in fall 2018 and covers Class 49.

Table 2.15
Core Component Completion Among ChalleNGe Graduates: Community Service, by Site (Classes 48 and 49)

	Residential Class 48			Residential Class 49		
Site	Service Hours Per Cadet	Dollar Value Per Hour	Total Community Service Contribution	Service Hours Per Cadet	Dollar Value Per Hour	Total Community Service Contribution
All Sites			$7,056,732			$7,317,824
AK	56	$27.45	$276,696	51	$27.45	$204,393
AR	81	$20.01	$167,092	71	$20.01	$124,453
CA-DC	56	$29.09	$171,049	62	$29.09	$227,251
CA-LA	44	$29.09	$220,153	52	$29.09	$291,947
CA-SL	52	$29.09	$280,276	54	$29.09	$308,019
D.C.	41	$39.45	$47,341	56	$39.45	$99,006
FL	60	$23.33	$226,646	77	$23.33	$283,833
GA-FG	50	$25.15	$254,726	47	$25.15	$160,697
GA-FS	65	$25.15	$364,701	57	$25.15	$320,383
GA-MV	45	$25.15	$86,013	51	$25.15	$137,163
HI-BP	116	$25.40	$285,801	98	$25.40	$262,270
HI-HI	126	$25.40	$198,425	120	$25.40	$204,216
ID	46	$21.83	$103,527	46	$21.83	$115,104
IL	61	$26.02	$195,880	68	$26.02	$245,859
IN	40	$23.73	$98,717	40	$23.73	$92,072
KY-FK	58	$21.17	$118,161	55	$21.17	$84,068
KY-HN	67	$21.17	$109,216	72	$21.17	$124,988
LA-CB	44	$22.30	$199,689	46	$22.30	$254,945
LA-CM	56	$22.30	$259,051	50	$22.30	$236,650

Table 2.15—Continued

Site	Residential Class 48			Residential Class 49		
	Service Hours Per Cadet	Dollar Value Per Hour	Total Community Service Contribution	Service Hours Per Cadet	Dollar Value Per Hour	Total Community Service Contribution
LA-GL	54	$22.30	$305,867	51	$22.30	$243,382
MD	52	$27.50	$143,633	51	$27.50	$142,606
MI	63	$23.91	$174,915	63	$23.91	$178,943
MS	73	$19.81	$244,962	65	$19.81	$255,838
MT	60	$22.42	$107,713	58	$22.42	$104,526
NC-NL	50	$23.41	$110,308	90	$23.41	$207,424
NC-S	100	$23.41	$206,214	83	$23.41	$204,510
NJ	48	$28.32	$111,235	48	$28.32	$81,137
NM	80	$20.58	$152,827	45	$20.58	$98,588
OK	60	$22.18	$135,031	49	$22.18	$122,899
OR	92	$24.89	$299,974	93	$24.89	$284,717
PR	70	$12.71	$195,734	50	$12.71	$137,268
SC	40	$22.22	$99,546	40	$22.22	$91,546
TN	N/A	N/A	N/A	43	$21.98	$21,738
TX-E	48	$24.64	$93,240	70	$24.64	$95,135
TX-W	40	$24.64	$40,410	40	$24.64	$60,122
VA	103	$26.75	$275,525	157	$26.75	$487,171
WA	65	$30.46	$283,126	67	$30.46	$287,756
WI	59	$24.00	$130,272	68	$24.00	$176,256
WV	63	$21.10	$200,724	62	$21.10	$200,155
WY	41	$23.17	$82,317	42	$23.17	$58,790

NOTE: Information in this table was reported by the sites in fall 2018 and covers Classes 48 and 49, which began in 2017. The value of community service is calculated using published figures at the state level for 2015, which are available online (Independent Sector, 2016). The value of community service was calculated in the same manner in the previous three annual reports (Wenger, Constant, and Cottrell, 2018; Wenger et al., 2017; National Guard Youth ChalleNGe, 2015).

N/A = not available.

Table 2.16
Residential Phase Performance Among ChalleNGe Graduates: Physical Fitness, by Site (Class 48)

Site	Average Number of Push-Ups		Average 1-Mile Run Time	
	Initial	Final	Initial	Final
All Sites	26.1	43.3	10:14	08:19
AK	31.5	42.9	09:54	08:04
AR	31.4	47.5	10:11	08:01
CA-DC	31.2	49.7	09:12	07:58
CA-LA	20.2	51.6	09:23	07:34
CA-SL	21.1	40.1	09:54	07:39
D.C.	29.9	33.4	13:11	11:44
FL	18.6	31.9	10:34	08:23
GA-FG	30.0	42.0	09:36	09:21
GA-FS	*	*	09:48	08:41
GA-MV	49.3	54.2	06:16	06:29
HI-BP	31.4	55.5	11:07	08:59
HI-HI	33.3	60.8	08:33	07:42
ID	20.1	41.3	09:58	07:53
IL	20.3	45.3	11:20	07:50
IN	*	*	10:45	08:09
KY-FK	27.1	50.2	10:59	09:46
KY-HN	34.4	*	09:27	*
LA-CB	25.4	36.5	09:29	10:52
LA-CM	21.1	35.2	10:43	08:30
LA-GL	24.4	48.4	10:00	08:20
MD	22.9	46.5	11:09	08:59
MI	33.8	51.6	09:31	08:04
MS	25.3	50.0	11:23	07:50
MT	*	*	12:09	08:05
NC-NL	17.0	34.2	13:02	11:01
NC-S	26.2	42.2	10:34	07:52

Table 2.16—Continued

Site	Average Number of Push-Ups		Average 1-Mile Run Time	
	Initial	Final	Initial	Final
NJ	29.6	55.9	11:03	08:34
NM	33.2	55.1	09:56	06:37
OK	38.5	42.1	10:55	09:13
OR	17.6	29.7	09:56	07:22
PR	*	*	*	*
SC	*	*	09:49	08:00
TN	N/A	N/A	N/A	N/A
TX-E	7.0[a]	7.9[a]	08:59	08:37
TX-W	7.4[a]	9.8[a]	10:19	09:23
VA	25.8	35.3	10:52	09:52
WA	25.4	40.1	10:40	07:39
WI	15.2	24.6	10:04	07:24
WV	*	*	10:12	07:30
WY	31.4	38.2	09:10	07:29

NOTE: Information in this table was reported by the sites in fall 2018 and covers Class 48.

* Did not report.

[a] Pull-ups; site does not collect data on push-ups.

N/A = not available.

Table 2.17
Residential Phase Performance Among ChalleNGe Graduates: Physical Fitness, by Site (Class 49)

Site	Average Number of Push-Ups		Average 1-Mile Run Time	
	Initial	Final	Initial	Final
All Sites	24.7	40.9	10:10	08:32
AK	22.9	39.2	09:58	08:17
AR	18.2	36.9	11:31	10:50
CA-DC	26.1	34.4	10:03	08:34
CA-LA	24.3	51.4	09:23	07:31
CA-SL	25.7	40.6	08:44	07:09

Table 2.17—Continued

Site	Average Number of Push-Ups		Average 1-Mile Run Time	
	Initial	Final	Initial	Final
D.C.	31.5	37.7	11:41	11:08
FL	16.5	31.0	10:53	08:25
GA-FG	31.5	44.6	09:47	09:18
GA-FS	29.0	46.1	09:51	08:19
GA-MV	43.1	50.7	08:38	07:54
HI-BP	23.5	50.6	11:08	08:27
HI-HI	37.6	50.1	08:34	07:16
ID	19.0	43.6	09:38	08:03
IL	20.2	48.4	10:26	09:11
IN	24.4	35.6	10:47	08:47
KY-FK	19.3	35.7	13:08	10:02
KY-HN	28.6	44.6	09:57	09:13
LA-CB	22.4	34.7	09:24	08:53
LA-CM	24.0	35.0	09:42	08:03
LA-GL	27.0	38.3	11:01	10:25
MD	23.9	33.1	10:51	09:36
MI	28.2	49.6	09:21	08:13
MS	22.6	48.3	11:59	07:59
MT	*	*	10:15	08:09
NC-NL	22.5	37.1	13:00	10:59
NC-S	15.4	34.4	11:15	08:28
NJ	29.5	46.0	10:36	11:42
NM	21.5	43.8	09:43	06:47
OK	28.4	44.9	09:54	08:48
OR	29.7	46.4	09:43	07:22
PR	24.3	37.8	09:43	08:07
SC	*	*	10:02	07:58
TN	37.9	46.0	10:40	09:49

Table 2.17—Continued

Site	Average Number of Push-Ups		Average 1-Mile Run Time	
	Initial	Final	Initial	Final
TX-E	6.6[a]	6.1[a]	12:12	09:45
TX-W	8.5[a]	54.0[a]	10:29	08:09
VA	25.9	42.8	09:28	08:32
WA	26.0	36.3	10:00	08:09
WI	10.6	25.7	09:14	07:50
WV	*	*	09:35	07:46
WY	18.7	25.1	10:46	09:02

NOTE: Information in this table was reported by the sites in the fall of 2018 and covers Class 49.

* Did not report.

[a] Pull-ups; site does not collect data on push-ups (TX-W switched to push-ups for its final assessment of Class 49).

Much of the information in this section has been reported consistently across multiple annual reports. On many metrics (e.g., test score gains, graduation rates), programs appear to be performing much as they did in the past. However, these tables also amply demonstrate the substantial variation that exists across programs. Some of this variation is related to program size, but other metrics, such as test scores and credentials awarded, are not obviously related to program size. And comparing some of the tables earlier in this chapter with information in past reports suggests that there are trends in the overall number of cadets in several programs and, perhaps, in the graduation rates of cadets and other metrics.

In the following section, we present a detailed analysis of a key education-related measure: performance on the TABE. It is important to note that this analysis includes only cadets who completed the program. This analysis is followed by information on site characteristics, program staffing, and cadet placements. In the last section of this chapter, we examine some trends across time.

Tests of Adult Basic Education Scores

The TABE is a standardized test with subtests that focus on reading, language arts, and math. The TABE is most commonly used in adult basic and secondary education programs.[1] At a minimum, ChalleNGe cadets take the TABE at the beginning of the

---

1   For more information about TABE and the common uses of the test, see U.S. Department of Education, Office of Vocational and Adult Education, Division of Adult Education and Literacy, 2016. For more informa-

program and again at the end of the Residential Phase; some sites also use the TABE more extensively to track progress during the course of the Residential Phase.[2]

Grade-equivalent scores on the TABE can be linked to outcomes of interest; for example, a grade-equivalent score of 9.0 is associated with a 70 percent passing rate on the reading, language arts, and math computation sections of the GED, while a grade-equivalent score of 11.0 is associated with an 85 percent passing rate on the same tests.[3] We report and analyze TABE information in terms of key grade-equivalent scores, and we requested that sites report scores for math, reading, and the TABE "total battery."[4] Figure 2.1 documents changes in the total battery of TABE scores over the Residential Phase of the ChalleNGe program. At the beginning of the Residential Phase, about three-quarters of cadets scored at the eighth-grade level or lower. However, by the end of the Residential Phase, more than half of cadets scored at or above the ninth-grade level. This suggests substantial progress and indicates that many of these cadets are quite likely to pass the GED test. Figure 2.1 summarizes the information found in Tables 2.11 and 2.12. Note that all TABE-related figures in this section are calculated using only graduates to ensure that we include the same group of cadets in each class who completed the pretest and the posttest.

Figure 2.2 presents similar information, but by subject area. Figure 2.2 shows that scores on the TABE reading test are higher than scores on the math tests; this is true both at the beginning and end of the Residential Phase. At the end of the Residential Phase, more than half of the cadets scored at or above the high school (ninth-grade) level on reading, while fewer than half achieved that benchmark on math. However, the proportion of cadets scoring at or above the ninth- and 11th-grade levels increased substantially between entry and graduation. Note that Figure 2.2 summarizes the information found in Tables 2.7–2.10.

Although all ChalleNGe sites focus on the program's eight core components and adhere to a variety of requirements and regulations, there is substantial variation across the sites, for many reasons. Examples include differences in the students who struggle in high school, the curriculum and services available in schools, the number and

---

tion about ChalleNGe's use of the TABE and the differences between grade equivalents and gain scores, see Wenger et al., 2017, and Wenger, Constant, and Cottrell, 2018.

[2] According to the data we collected in fall 2018, nine of the 40 ChalleNGe sites reported routinely administering the TABE more than twice during the Residential Phase.

[3] For example, see National Reporting System for Adult Education, undated; Comprehensive Adult Student Assessment System, 2016; Comprehensive Adult Student Assessment System, 2003; West Virginia Department of Education, undated; Olsen, 2009; Wenger, McHugh, and Houck, 2006; and Wenger et al., 2017.

[4] The TABE total battery is formed from the scores on reading, language arts, math computation, and applied math. In an effort to minimize burden on the sites, we collected data on the total battery, reading, and math computation tests only; note that these subject tests are the ones that have been found to be predictive of performance on the GED test. In past data collection efforts, we also collected data on the language arts subtest, but we found that the reading subtest provided very similar information.

Figure 2.1
Cadet Scores on TABE Total Battery at Beginning and End of ChalleNGe Residential Phase

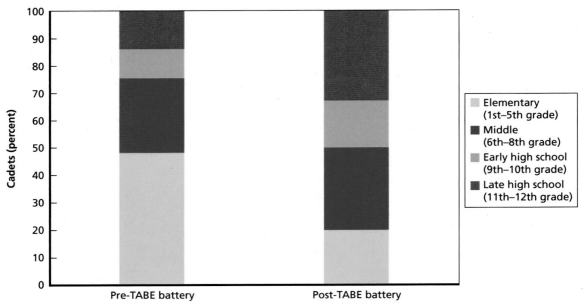

NOTE: This figure is based on information reported by the sites in fall 2018 and covers graduates from Classes 48 and 49.

Figure 2.2
Cadet Scores on TABE Subject Tests at Beginning and End of ChalleNGe Residential Phase

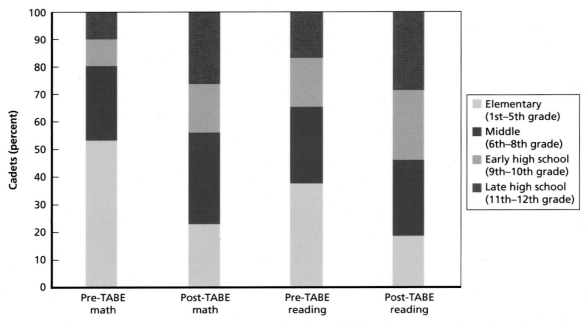

NOTE: This figure is based on information reported by the sites in fall 2018 and covers graduates from Classes 48 and 49.

type of alternative programs available in the state, the extent to which ChalleNGe is well known in the state, and the background and experience of the instructors and the cadre who oversee the cadets. Any or all of these differences could influence the number of applicants, the graduation rate, or the academic progress cadets make in their programs.

Finally, as discussed above, different ChalleNGe programs provide different academic credentials: Some programs provide a GED or HiSET certificate, some provide high school credits, and some give cadets the opportunity to complete high school and receive a diploma. In fact, many programs are hybrids; they provide different credentials to differently prepared cadets. Nonetheless, the majority of graduates at most programs receive a single type of credential. Based on the information collected from the sites in fall 2018, we divided the programs into two groups: programs primarily granting GED or HiSET certificates and those primarily granting high school credits or diplomas. There are many other ways in which we could have divided these programs; we selected this division based on the hypothesis that programs focusing on the GED or HiSET might have a different curriculum and possibly even different entrance requirements *or* differently prepared cadets. Figure 2.3 shows pre- and post-TABE total battery scores for cadets who completed ChalleNGe, separated by those who attended sites that primarily grant GEDs or HiSETs and those who attended sites primarily granting high school credits or diplomas.

Figure 2.3 demonstrates that graduates at sites granting high school credits or diplomas are slightly *more* likely to score at the elementary or middle school level when entering the ChalleNGe program. Scores for cadets in both types of programs are very similar at the end of the Residential Phase, although cadets at sites granting high school credits or diplomas are still more likely to score at the elementary or middle school level and less likely to score at the upper high school level.[5] The pattern is similar for math and reading subscores. These results suggest that cadets entering programs granting GED or HiSET certificates are, on average, slightly better prepared for coursework than cadets entering programs that grant high school credits or diplomas (based on TABE scores). Although graduates of programs granting high school credits or diplomas make substantial progress—and have higher test score gains than graduates of GED or HiSET programs—even at graduation there is no indication that cadets at programs granting high school credits or diplomas hold an advantage in terms of test scores.[6] The difference in graduation rates also may be relevant—cadets who enter programs that grant high school credits or diplomas are slightly more likely

---

[5] In each case, the noted differences are statistically significant; t-tests indicate that such differences would occur by chance less than one time out of 100.

[6] The TABE is only one measure of progress. Although staff are likely motivated to produce the maximum possible gain on the TABE, it is less clear how we should expect TABE gains to vary by credential awarded. To the extent that the TABE is similar to the GED or HiSET, practicing for these tests could increase TABE scores while at least some of the coursework for high school credits may be less closely related to the TABE. Nonetheless, our

Figure 2.3
Cadet Scores on TABE Battery Tests at Beginning and End of ChalleNGe Residential Phase, by Credentials Awarded

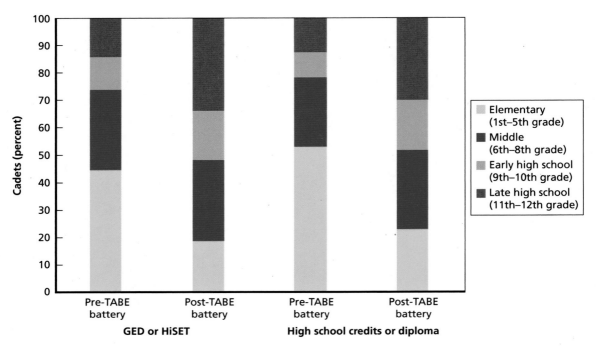

NOTE: This figure is based on information reported by the sites in fall 2018 and covers graduates from Classes 48 and 49.

to graduate than cadets in GED or HiSET programs (see Figure 2.14 and the accompanying text). This interaction between credentials granted and graduation rates is worthy of additional attention; program-level analyses planned for 2019 will include a focus on this interaction.

Site Characteristics

Along with differences in the credentials awarded, sites differ on numerous other factors as well. As a first step toward determining the relationship between these differences and cadet success, we collected information on several program-related policies, including admissions factors, disciplinary policies, policies related to placement of cadets in platoons, additional activities provided to cadets, job placement supports, and college placement supports.

As shown in Figure 2.4, sites reported incorporating a variety of factors in their admissions policies. Most sites incorporated behavioral or mental health consider-

results suggest that courses awarding high school credits or credentials do not begin their programs with cadets at substantially higher levels of achievement compared with those in other programs.

ations in their admissions considerations, as well as juvenile justice records and physical health. Few reported using TABE scores in their admissions considerations.

Sites also reported using a variety of policies to encourage positive behavior. The vast majority of sites reported the use of usual privileges and CAPE (physical exercise) to enforce disciplinary policies, as well as the promise of additional privileges on campus to cadets who follow rules and policies (Figure 2.5). More than half of the sites reported providing additional privileges off campus (such as field trips) to cadets who follow rules and policies. These options seem to capture much of the range of policies in place; only a minority of sites reported using other types of policies.

Sites also factored in certain considerations when assigning cadets to platoons (Figure 2.6). Most sites did not place cadets from the same family in the same platoon. A handful of sites reported that they factored in age, but around half of the sites reported taking a cadet's hometown, friendships, and gang membership into consideration.

On our site visits, we also encountered variety in terms of the activities and ceremonies available to cadets. To gain a better understanding of this variation across ChalleNGe program sites, we collected information on the specific activities and ceremonies available to cadets at each site (see Figure 2.7).

While the majority of sites provided most of the activities and ceremonies listed in Figure 2.7, other activities were less common. Most sites offered student government, a

Figure 2.4
Admissions Factors

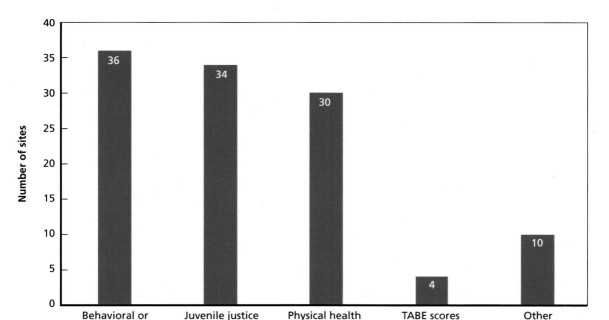

NOTE: This figure is based on information reported by 40 sites in fall 2018.

Figure 2.5
Policies to Encourage Positive Behavior

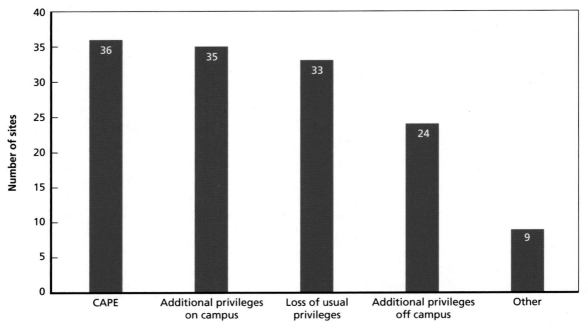

NOTE: This figure is based on information reported by 40 sites in fall 2018.

Figure 2.6
Factors Taken into Consideration When Placing Cadets in Platoons

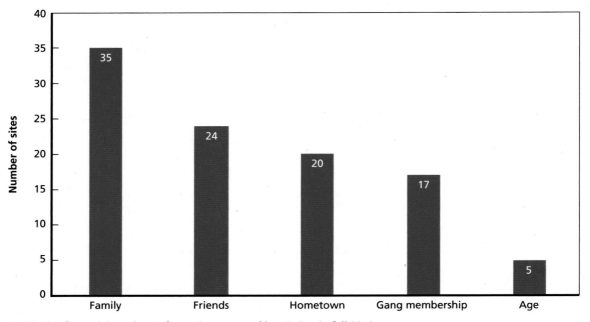

NOTE: This figure is based on information reported by 40 sites in fall 2018.

graduation ceremony, a color guard, and a library. More than half of the sites held an awards banquet. Some sites sponsored a prom for cadets, and some sponsored running clubs. But our list of activities and ceremonies failed to include many of the options actually offered. In our site visits, we heard about a variety of additional sports, a drill team, a drum corps, and clubs for such activities as chess, Latin, math, and various arts (these activities are categorized as *Other* in Figure 2.7). These types of ceremonies and activities could be related to the probability of graduating from ChalleNGe; we will explore these potential relationships in program-level analyses in 2019.

Most sites provided a range of activities to assist cadets with job placement after completing the ChalleNGe program. Common activities included mock interviews, assistance with job seeking and completing job applications, career tests, and various other career exploration activities. We did not find significant variation in these activities across sites. About half of the sites offered cadets opportunities to gain work experience (Figure 2.8).

College placement supports were also common across sites. Most sites help cadets compare two-year community college and four-year college or university options. The vast majority of sites helped cadets complete applications and provided access to relevant information. Fewer sites, although still a majority, held or participated in college

Figure 2.7
Activities and Ceremonies Provided to Cadets

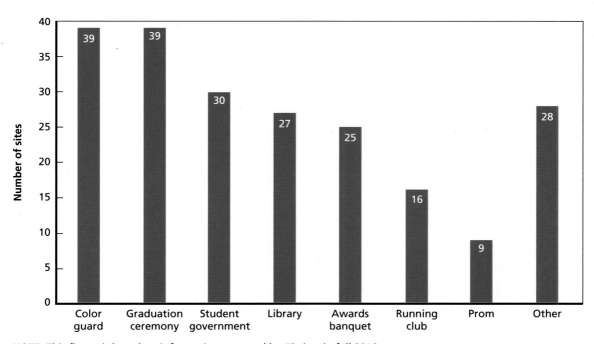

NOTE: This figure is based on information reported by 40 sites in fall 2018.

fairs or held information sessions on postsecondary education options and resources for cadets, parents, and mentors (Figure 2.9).

Staffing

We collected information on the number of staff by position, as well as the number of staff by position newly hired in the last 12 months. Although this information could be an indicator of excessive turnover, it is important to keep it in the context of other changes that may occur with programs. For example, changes in leadership may result in staff leaving and new staff arriving. Newer programs may have less stability in their staff because it may take time for the program to develop an identity and select staff with the right fit. Finally, staff turnover can be the result of natural occurrences as well, such as retirement or a tendency for short tenure for some positions. Figures 2.10 to 2.12 provide information on the number of sites reporting that 25 percent or more of site staff were newly hired in the last 12 months. Sites were reporting on their 2017–2018 year and are grouped by year of establishment.

At about half of all sites, at least 25 percent of total staff were hired in the last 12 months (Panel A, Figure 2.10). Among the older programs (established prior to

Figure 2.8
Job Placement Supports Provided to Cadets

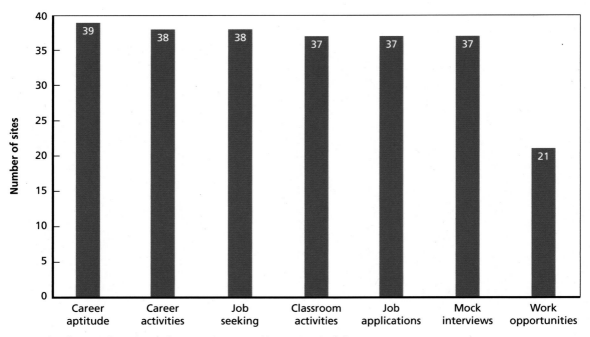

NOTE: This figure is based on information reported by 40 sites in fall 2018.

Figure 2.9
College Placement Supports Provided to Cadets

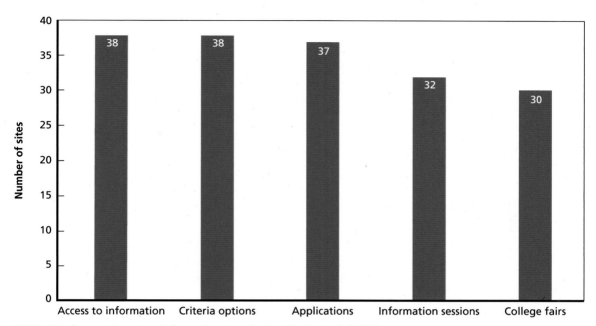

NOTE: This figure is based on information reported by 40 sites in fall 2018.

2000), a relatively large share of recruiters (Panel B, Figure 2.10) and cadre (Panel A, Figure 2.12) were newly hired within the last 12 months. With the exception of these positions, however, staff turnover at the most established programs was markedly low, and substantially lower than turnover at newer programs. Half or more of the newer programs reported that at least 25 percent of their instructors, case managers, and cadre were newly hired in the last 12 months. Figures 2.10–2.12 suggest considerable turnover among cadre and, to a lesser extent, among recruiters. Turnover in other positions was low at long-established sites but markedly higher at newer sites (even those that have been in existence for a decade or more). Staff turnover is expected and could be positive in some cases. However, excessive turnover may result in negative program- and cadet-level outcomes.

Program Completion

Most of our analyses in this chapter focus on the cadets who successfully complete, or *graduate* from, the ChalleNGe program. This analytic decision was driven by our focus on changes over the course of the program (in TABE scores and physical fitness) and on outcomes after the program (such as placement, discussed below). The measures required for these analyses do not exist for cadets who leave a site prior to com-

Figure 2.10
Number of Sites with at Least 25 Percent of Staff Hired in the Last 12 Months:
Total Staff, Recruiters

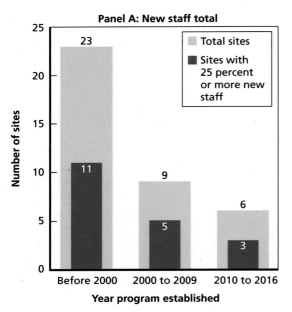

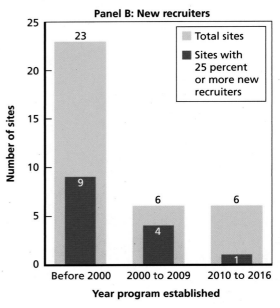

NOTE: This figure is based on information reported by 38 sites for total new staff and 35 sites for new recruiters in fall 2018.

Figure 2.11
Number of Sites with at Least 25 Percent of Staff Hired in the Last 12 Months:
Instructors, Case Managers

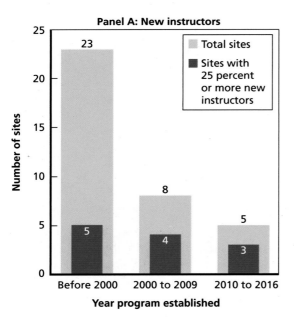

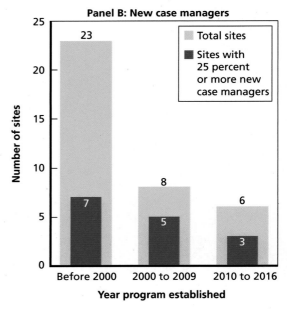

NOTE: This figure is based on information reported by 36 sites for new instructors and 37 sites for new case managers in fall 2018.

Figure 2.12
Number of Sites with at Least 25 Percent of Staff Hired in the Last 12 Months:
Cadre, Administrators

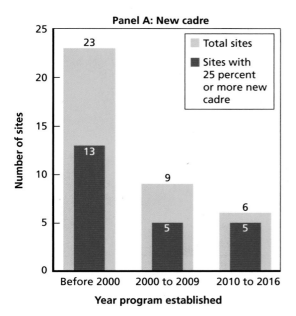
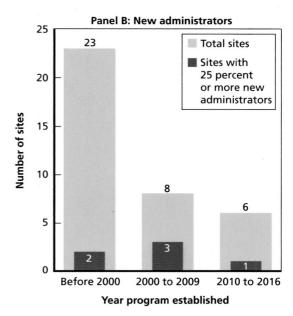

NOTE: This figure is based on information reported by 38 sites for new cadre and 37 sites for new administrators in fall 2018.

pleting the ChalleNGe program. But graduation rates vary across sites, and they could be viewed as another key site-level outcome. Also, graduation rates are likely related to other program characteristics. For these reasons, we examine graduation rates in detail in this section. We present cadet-level analyses that compare the characteristics of graduates with those of nongraduates, and we explore the variation of graduation rates by site characteristics (such as credentials awarded).

Of youth who enroll in the ChalleNGe program, roughly 25 percent leave prior to graduation.[7] Past analyses suggest that female cadets, cadets who are at least 17 years of age at entry, cadets with higher TABE scores, cadets from areas with lower levels of poverty, and cadets who live further from their ChalleNGe site are more likely than other cadets to graduate.[8] Additionally, site-level characteristics appear to explain graduation rates; for example, sites at which staff rated the level of militarization as *high*

---

[7]   Some cadets request to leave the program; in other cases, the final decision is made by program staff. This distinction may or may not be meaningful: Cadets who wish to leave may choose to deliberately violate rules rather than follow the multistep policy required of those who request to leave. Conversations with program staff indicate that staff generally expend considerable effort to retain cadets.

[8]   These results were produced by regression models holding constant the site, year, and class (first or second class of the year); the data were from 1999–2006. Overall graduation rates were somewhat lower during that period. See Wenger et al., 2008.

Figure 2.13
Graduation Rates, by Cadet Characteristics

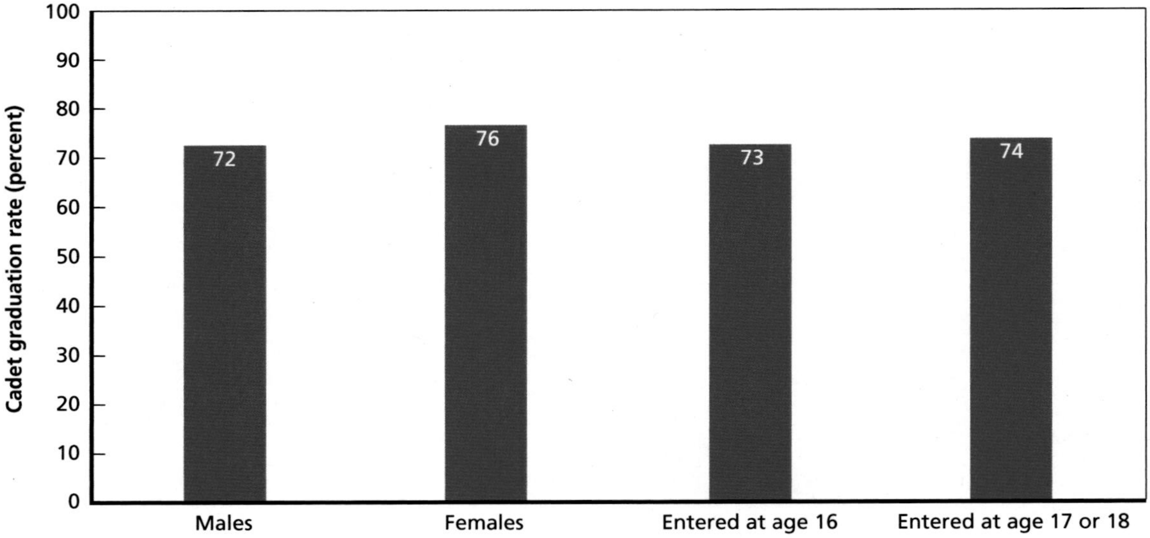

NOTE: This figure is based on information reported by the sites in fall 2018 and covers graduates from Classes 48 and 49. The graduation rates above were calculated using cadet-level data. The difference between males and females is statistically significant (one-tailed t-test statistic < 0.0001) and thus is extremely unlikely to have occurred by chance. The difference by age at entry is marginally significant (one-tailed t-test statistic = 0.06) and thus would be expected to occur by chance only six times in 100.

(versus *medium* or *low*) had higher graduation rates, holding constant other factors (Wenger et al., 2008).

Next, we present descriptive statistics comparing graduation rates of cadets with different individual characteristics. Figure 2.13 includes the graduation rate (calculated from cadet-level data) of male versus female cadets and of cadets by age at entry. As was the case in prior analyses, female cadets graduated at a higher rate than male cadets. This could represent some form of selection—for example, girls who are interested in ChalleNGe may be more motivated than boys. The difference by age at entry is also present, although the effect is smaller (and only marginally significant). In general, these findings are consistent with earlier research, and they suggest that individual characteristics explain some of the likelihood of success. They also suggest that programs should expect higher graduation rates if they admit more girls or older cadets.

We also examined graduation rates by credential awarded. Figure 2.14 demonstrates that graduation rates were actually slightly higher at programs awarding high school credits or diplomas. Recall from Figure 2.3 that TABE test scores at the beginning of the Residential Phase were slightly *lower* at sites offering high school credits or diplomas, suggesting that cadets at sites awarding high school credits or diplomas do not enter ChalleNGe with an academic advantage. Although there are many other factors that influence graduation rates, Figure 2.14 indicates that cadets at the sites award-

Figure 2.14
Graduation Rates, by Credentials Awarded

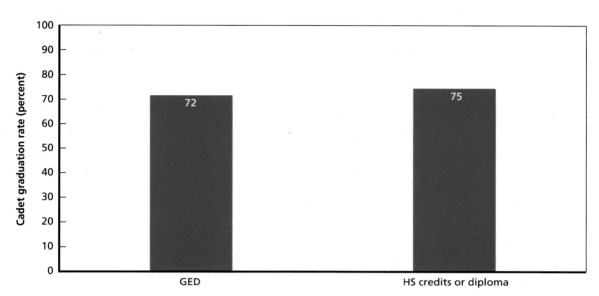

NOTE: This figure is based on information reported by the sites in fall 2018 and covers graduates from Classes 48 and 49. See Table A.1 for a list of programs by credential. The graduation rates above were calculated using cadet-level data. The difference is statistically significant (one-tailed t-test statistic < 0.0001) and thus is extremely unlikely to have occurred by chance.

ing high school credits or diplomas are at least as likely to complete ChalleNGe as cadets at sites awarding GEDs. In other words, awarding high school credits or diplomas does not seem to come at a cost of lower graduation rates. It is possible that the presence of these credentials serves as a motivating factor for some cadets who might not otherwise complete the program; indeed, the presence of these credentials and the increase in the programs offering them could be one reason that current graduation rates are higher than those found in data from the early 2000s.[9]

Along with cadets' personal characteristics (such as gender or age at entry) and credentials offered, other site-level characteristics could be related to graduation rates. Staff turnover is an example of such a site-level characteristic. Turnover may be disruptive to cadets, and training new staff members may present a substantial burden to current staff. Turnover could influence graduation rates through one or both of these routes, or through other routes.

We presented information detailing the number of sites with substantial staff turnover (see Figures 2.10–2.12). Given the frequent interactions between cadre and cadets, next we focus on cadre turnover. Many programs had relatively high levels of

---

[9]   Of course, other factors could influence this relationship—for example, cadets at sites awarding high school credits or diplomas could have higher overall levels of motivation. But, again, our analyses suggest that awarding these credentials does not involve accepting lower graduation rates.

turnover among cadre. Figure 2.12 indicates that 23 of the ChalleNGe sites reported turnover rates of at least 25 percent among cadre, and both newer and more established programs had high turnover rates. When we compared graduation rates across programs with different levels of staff turnover, we found that graduation rates were slightly lower at programs with higher reported cadre turnover in the past year (Figure 2.15).[10]

It is important to note that these descriptive analyses shown do not take into consideration the site-level factors that could be driving graduation outcomes. To simultaneously consider the multiple individual and site-level drivers potentially associated with graduation outcomes, we also conducted multivariate regression analyses on some of the outcomes (not depicted here). The regressions included individual and site-level factors, and the findings were quite consistent with those shown here.

These results have implications for the ChalleNGe program. Graduation rates appear to be higher than in past decades. Offering additional credentials is not associated with lower graduation rates. However, younger cadets and male cadets are less likely than others to complete the program; this finding could imply a need for additional support resources. Sites with higher cadre turnover rates have somewhat lower cadet graduation rates, which could be a measure of the disruption caused by staff

Figure 2.15
Graduation Rates, by Cadre Turnover

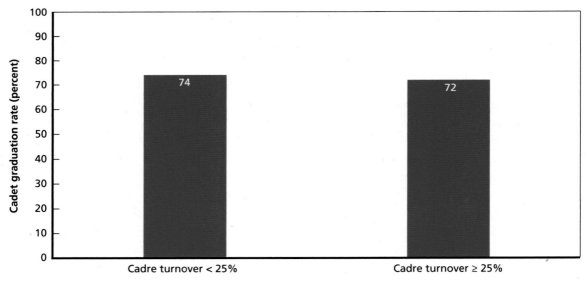

NOTE: This figure is based on information reported by the sites in fall 2018 and covers graduates from Classes 48 and 49. The graduation rates above were calculated using cadet-level data. The difference is statistically significant (one-tailed t-test statistic = 0.03) and thus is unlikely to have occurred by chance.

---

[10] Programs that have opened recently would, of course, have hired staff recently. We exclude these programs from Figure 2.15.

turnover. This finding suggests that decreasing staff turnover could have positive implications for cadets.

Sites seeking to improve their graduation rates could consider a rigorous analysis of the cadets who do not complete the program. Such an analysis would optimally include a comparison of the measurable characteristics of cadets, as well as qualitative information (such as the reason[s] for departure). Such analyses could help sites as they seek to increase their impact on cadets and communities. Indeed, as sites seek to measure the longer-term impacts of the ChalleNGe program, a careful tracking or benchmarking of graduation rates over time should be included to ensure that long-term outcomes do not come at the cost of program completion.

Placement

During the course of the ChalleNGe program, all cadets develop a post-ChalleNGe plan. Sites use a specific form, referred to as a Post-Residential Action Plan (P-RAP), as a tool to assist cadets in developing their plans. Plans can be quite detailed and can include additional education, searching for and obtaining a job, joining the military, or some combination of these options. From the perspective of the ChalleNGe program, a successful placement is defined as any one or combination of education, employment, or military participation. As was the case in past data collections, we requested and received information on the placement of recent graduates. At the time of our data collection, graduates of Class 49 (who generally entered ChalleNGe during the latter half of 2017) had, in most cases, completed the program within the prior 12 months. Therefore, we report placement information only at months one and six for these cadets. The placement information is only available for cadets who completed the program and graduated.

Figure 2.16 shows placements of Classes 48 and 49 at three points after graduation. Although nearly one-third of graduates did not have a known placement in the first month after graduation, this figure falls in later months: By months six and 12, at least three-quarters of graduates were listed as having a placement. Education and employment were the most common placements. Cadets were most likely to be enrolled in school in the first month after graduation; in later months, cadets were more likely to be employed. The proportion of cadets who reported military service increases over the months, as does the proportion who reported some combination of education, employment, and military service. In the first month after graduation, about 60 percent of cadets were enrolled in school, employed, or serving in the military (or some combination of the three). By months six and 12, roughly seven out of ten recent ChalleNGe graduates were involved in some combination of these activities. At each point, between 6 percent and 10 percent of cadets were volunteering or serving their communities in some manner or reported a similar activity (these activities are categorized as *Other* in Figure 2.16).

Note that Figure 2.16 does *not* include cadets whose records are missing placement information: Qualitative information gathered during our site visits suggests that some or perhaps even most of these graduates are likely placed, but the programs struggle to obtain placement information on them. One driver of this trend is the difficulty sites encounter when trying to maintain contact with mentors. As noted in Wenger et al. (2017), sites reported spending considerable time and effort trying to maintain contact with mentors. Despite these efforts, less than three-quarters of mentors were reporting information on cadet placements six months after graduation, and the proportion of mentors in contact with their cadets continued to fall in the next few months (Wenger et al., 2017). Among Classes 48 and 49, sites reported information on 80 percent of cadets at the six-month mark and 73 percent at the 12-month mark. These percentages are noteworthy: They imply that sites are able to obtain placement information on

Figure 2.16
Cadet Placements in Months 1, 6, and 12 After Graduation

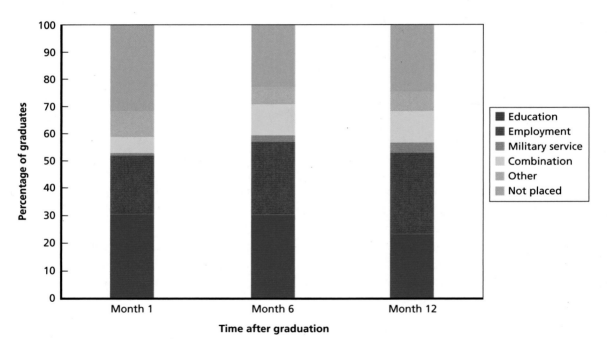

NOTE: This figure is based on information reported by the sites in fall 2018 and covers graduates from Classes 48 and 49. Twelve-month placements include only Class 48. "Other" placements include placements that are noted as volunteer or service to community positions, as well as placements that are recorded simply as "Other." In each month, there are a substantial number of records with no placement information due to difficulties in contacting cadets or mentors. Note that this figure classifies nonplacements in a somewhat different manner than our earlier reports (Wenger et al., 2017; Wenger, Constant, and Cottrell, 2018); therefore, placement rates reported here are not directly comparable to earlier rates. Rates reported here reflect all information collected from mentors or others and exclude cadets for whom no information is available.

some cadets despite losing contact with the cadets' mentors. This reporting indicates that sites continue to expend resources and effort to document cadet placements. Based on information gathered during site visits, we find that sites use multiple strategies to contact cadets and document placement. Additional mentor training or other outreach efforts could improve mentor contact rates. We plan to launch a pilot program with at least one site in 2019 to test potential interventions aimed at improving reporting rates.

Time Trends, 2015–2017

One focus of this project is to collect consistent, cadet-level data across time. We have begun this process and now have data on six classes (the classes, two per year, that began in 2015, 2016, and 2017). Such data are very useful not only for determining relationships between policies and cadet success but also for documenting trends over time. We will continue to collect this information throughout the project and will be able to include data on eight classes in our final report.

Next, Figures 2.17 and 2.18 present some time trends in terms of the following outcomes:

- the total number of cadets
- the number of graduates
- the number of cadets whose TABE total battery scores exceed the ninth-grade level at graduation
- the percentage of cadets who graduated
- the percentage of cadets whose TABE total battery scores exceed the ninth-grade level at graduation.

As shown in Figure 2.17, the number of cadets participating in ChalleNGe increased slightly over this period. This appears to reflect both an increase in the number of sites and a small increase in the average number of cadets per site. The total number of graduates also increased, while the graduation rate remained roughly constant (Figure 2.18). The number of cadets scoring at or above the ninth-grade level at graduation also increased, although this growth was very slight and was much smaller than the overall growth in graduates. The proportion scoring at or above the ninth-grade level on the TABE remained roughly constant.[11] In short, the number of participants and the number of graduates in the ChalleNGe program increased over the last six classes. The metrics on overall graduation rates and TABE scores suggest that the sites were able to increase the number of participants without a marked decrease in graduation rates or achievement, as measured by test scores.

---

[11]   In general, growth in TABE scores incorporating tests taken at the beginning and end of the Residential Phase across sites has been relatively constant over time.

Figure 2.17
Trends in Cadets, Graduates, and TABE Battery Scores over Time

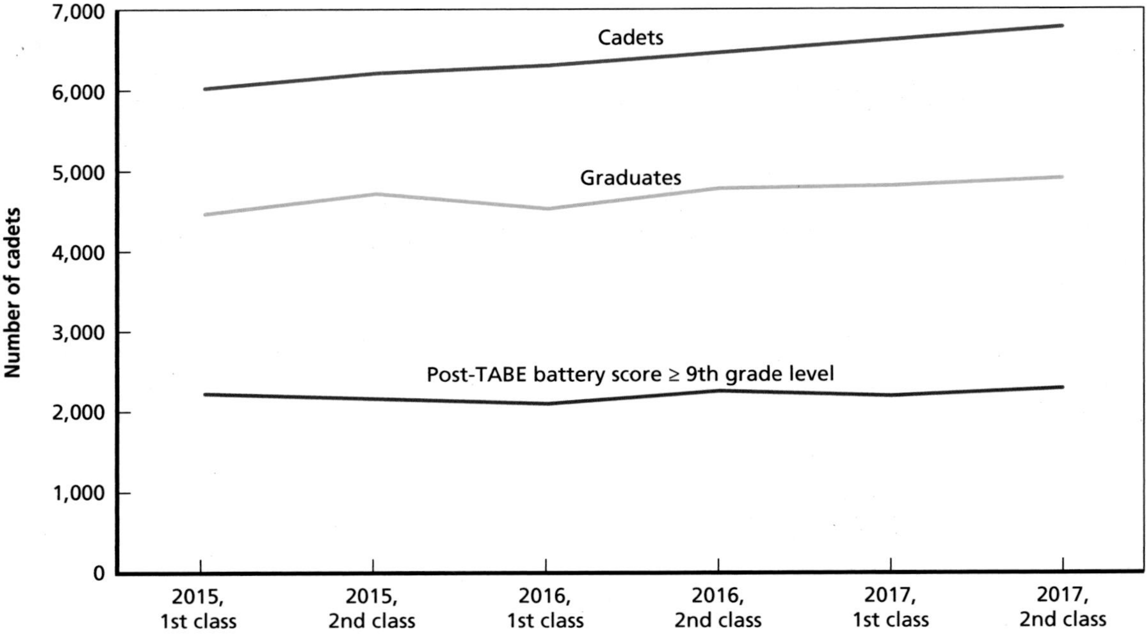

NOTE: Aggregated cadet-level data from each site; this figure is based on information reported by the sites in fall 2018 for classes 48 and 49, as well as information collected for previous reports (Wenger et al., 2017; Wenger, Constant, and Cottrell, 2018). The data included in this figure are not directly comparable with the information included in Figure 2.8 in Wenger, Constant, and Cottrell, 2018. In that report, we excluded data from Puerto Rico because the Puerto Rico site was disrupted by Hurricane Maria and was therefore unable to report data. However, the Puerto Rico site reported current and past data during fall 2018; therefore, this figure includes data from all sites.

Summary

In this chapter, we documented progress made across the ChalleNGe sites in 2017–2018, focusing on the two classes of cadets who entered a ChalleNGe program in 2017. The data indicate that cadets made considerable progress in terms of academic and nonacademic outcomes and that overall graduation rates have remained roughly constant. However, many of our analyses here are descriptive in nature, and it is unclear what drives many of the program differences we report in this chapter. In the next chapter, we describe our other analytic efforts to better understand how well ChalleNGe is meeting its goals. Some of these efforts focus on explaining program-level differences.

Figure 2.18
Trends in Graduation Rate and TABE Battery Scores over Time

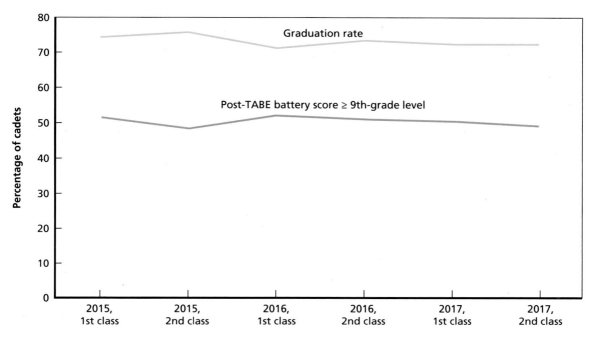

NOTE: Aggregated cadet-level data from each site; this figure is based on information reported by the sites in fall 2018 for classes 48 and 49, as well as information collected for previous reports (Wenger et al., 2017; Wenger, Constant, and Cottrell, 2018). The data included in this figure are not directly comparable with the information included in Figure 2.8 in Wenger, Constant, and Cottrell, 2018. In that report, we excluded data from Puerto Rico because the Puerto Rico site was disrupted by Hurricane Maria and was therefore unable to report data. However, the Puerto Rico site reported current and past data during fall 2018; therefore, this figure includes data from all sites.

CHAPTER THREE

Assisting ChalleNGe Through Benchmarking Youth Outcomes and Research on Mentoring

In this chapter, we discuss our framework for measuring the longer-term outcomes of the program. Much of our framework relies on two models (a TOC model and a logic model) that we developed in our 2017 report (Wenger et al., 2017). We include the logic model here, along with a brief discussion of its use to date in the project. We also discuss the implications for the longer-term metrics under development. We then describe the status of the ongoing site visits, which inform multiple aspects of the overall study objective. Analysis of the site visits, once they are all completed, will generate the themes we explore in our final research report published at the end of the study. These separate analytic efforts were the result of some of our early site visits, in which program staff described the challenges they faced in assessing program effectiveness. The site visits further add context to our interpretation of the data we collect annually from program sites to support the ChalleNGe program's yearly report to Congress.

Following our discussion of the logic model and the status of the site visits, we delve into preliminary findings from two of the analytic efforts currently underway:

- developing benchmarks for at-risk youth that will enable program sites to compare the education, employment, and social outcomes of their participants against youth in the broader population with the same profile
- examining the literature to identify best practices and approaches to support the development of more-robust youth-initiated mentoring (YIM).

To augment the literature review, we share early findings from our analysis of qualitative data on mentoring that we collected during our site visits.

Finally, we also describe a variety of other analytic efforts to support the development of longer-term metrics. These efforts include developing a pilot to analyze survey data on graduates and determine the efficacy of such data to track longer-term outcomes, well past the Post-Residential Phase; implementing improvements to mentor training and supports during the Residential and Post-Residential Phases; examining approaches to providing career and technical education (CTE) across program sites; and examining the association between program characteristics and participant outcomes. Collectively, as in the benchmarking and mentoring studies, these analytic efforts draw on multiple data sources, including information collected annually from the program sites, qualitative data collected from site visits, and other extant data sources, such as publicly available data from nationally representative surveys.

All of these analytic efforts were developed in response to key issues that we identified during site visits and in our direct interactions with program site leadership and staff, as well as input from our sponsor. These by no means represent *all* the issues facing program sites: our choice of which issues to address was based on a combination of interest in the key issue, availability of expertise on the team to address the issue, and feasibility of the analytic effort in terms of access to data and program site ability to support the research where needed (as in the case of the pilot). Budget and timeline constraints were also key considerations.

Logic Model

As noted above, we developed a program logic model—which we introduced in the first annual report (Wenger et al., 2017)—to assist in our development of longer-term metrics. The model delineates the inputs, processes or activities, expected outputs, and desired outcomes of the ChalleNGe program (for more detail, see Shakman and Rodriguez, 2015).

Program logic models are a useful way of specifying the reasoning behind program structure and activities, as well as how those activities are connected to expected program results (Knowlton and Phillips, 2009). They are used to illustrate how program resources, activities, services (inputs), and direct products of services (outputs) are designed to produce short-term, medium-term, and long-term outcomes. Logic models also identify broader community effects that should result from program activities and services (Knowlton and Phillips, 2009). In this way, these models can communicate how a program contributes not only to the specific needs and outcomes of participants but also to the broader community and society at large. Program logic models also serve as a blueprint for evaluating how effectively a program is meeting its expected goals.

The ChalleNGe logic model emphasizes the temporal aspects of the ChalleNGe program and its influence on participants, and it lays out expected results in detail (see Figure 3.1). The initial logic model was informed by a review of program documentation and annual reports and by site visits to two ChalleNGe locations (the Mountaineer ChalleNGe Academy in West Virginia and the Gillis Long ChalleNGe site in Louisiana). We have used the ChalleNGe logic model to clarify our thinking about the program's inputs, outputs, and outcomes. We also have used the model to communicate key aspects of the ChalleNGe program to a variety of stakeholders, including policymakers, program directors, program staff, and other researchers. In each case, the model has been a helpful communication tool. We will continue to refine the current program logic model and its uses in future reports.

At present, the ChalleNGe program sites focus primarily on collecting and reporting on metrics associated with the inputs, activities, and outputs. Metrics on short-term outcomes (achieved within one year) are collected and reported on, but the extent and consistency of reporting varies across program sites. Some of the outcome metrics included in Figure 3.1 (especially those listed as community outcomes) may be initially influenced by the program participants; for example, ChalleNGe graduates may vote and perform community service at higher levels than would be expected without program participation. However, it is also possible that the program will eventually have a broader influence on community-level outcomes.

Currently, program sites collect information on *graduates*, not on those who participated but did not complete the program (nongraduates). Although collecting information on all program participants (graduates and nongraduates) would be preferred, program sites currently have limited ability to do so because they rely on mentors to report information on graduates. Nongraduates are not formally assigned a mentor or tracked. Because sites collect contact information on all participants, sites could consider administering a survey to capture a sample of graduates and nongraduates. The efficacy of a survey to collect placement information is currently being explored as a pilot with one of the program sites (we provide an explanation of the pilot later in this chapter). One concern is that response rates are likely to be low among nongraduates without significant investment of survey administration resources. The treatment group in the MDRC study referenced in Chapter One included those admitted in its "intent-to-treat" design, which meant collecting information on both graduates and nongraduates (Bloom, Gardenhire-Crooks, and Mandsager, 2009; Millenky, Bloom, and Dillon, 2010; Millenky et al., 2011). However, this was a one-time study requiring significant investment of data collection resources that is difficult to replicate on an ongoing basis. Nonetheless, avenues to collect information on both graduates and nongraduates should continue to be explored to provide sites with a more complete picture of their performance.

The logic model has several implications for the ChalleNGe program. As we noted above, ChalleNGe's mission is to produce graduates who are successful, productive citizens in the years after they complete the program. The research that established the effectiveness of ChalleNGe on job performance and earnings, and the cost-benefit calculations associated with that research, focused on longer-term outcomes, which we can refer to as *impacts* (see Chapter One of this report, as well as Wenger et al., 2017). Currently, program sites focus on collecting outputs, which are considered measures of performance, not effectiveness. In order to measure effectiveness, program sites will need to focus more on collecting outcomes in order to determine the extent to which ChalleNGe is meeting its stated *mission*. Much of the information presented in Chapter Two focuses on inputs and outputs (the left-hand side of the logic model) and, to a

Figure 3.1
Program Logic Model Describing the National Guard Youth ChalleNGe Program

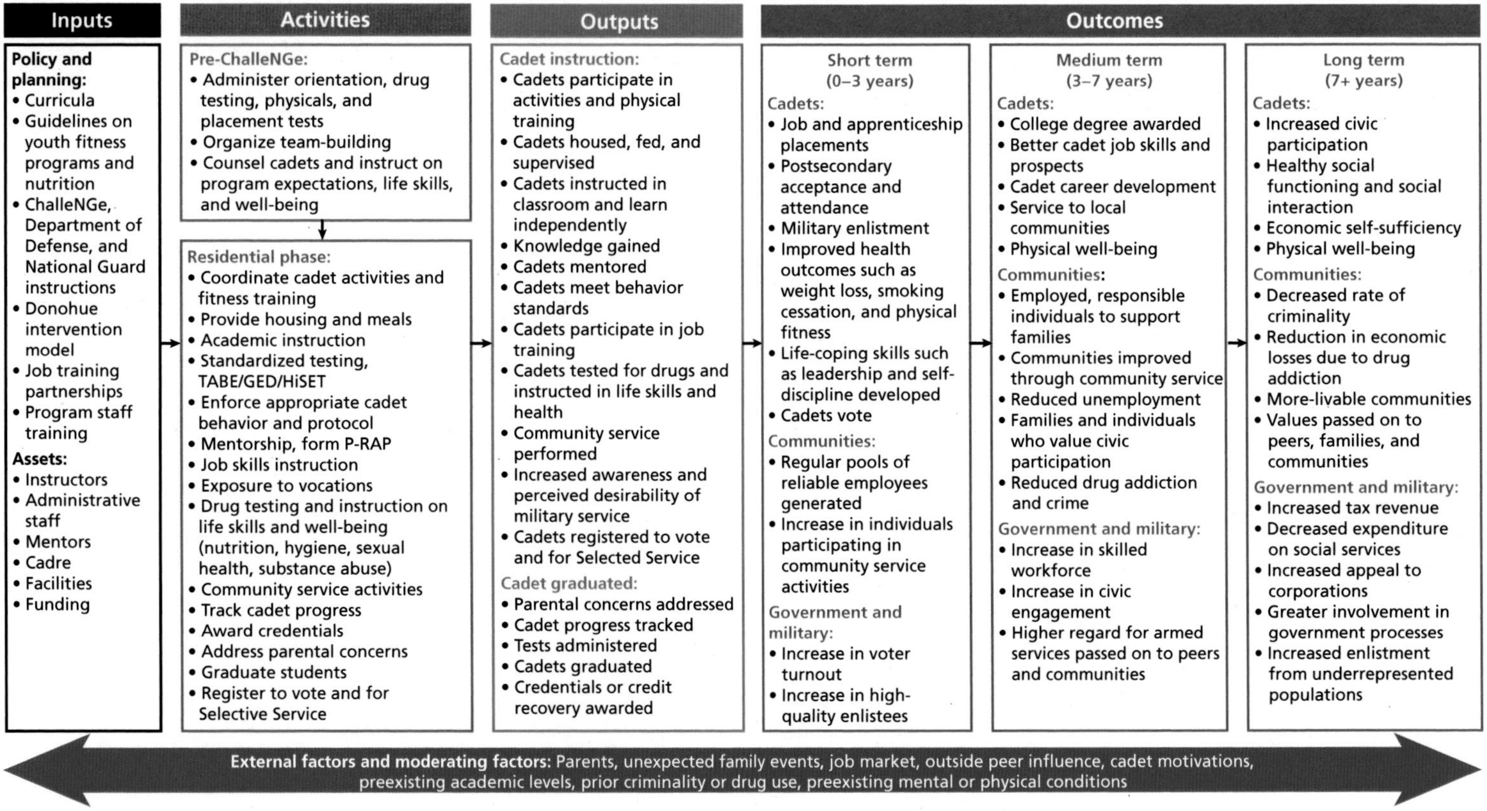

Inputs	Activities	Outputs	Outcomes		
			Short term (0–3 years)	Medium term (3–7 years)	Long term (7+ years)

Inputs

Policy and planning:
- Curricula
- Guidelines on youth fitness programs and nutrition
- ChalleNGe, Department of Defense, and National Guard instructions
- Donohue intervention model
- Job training partnerships
- Program staff training

Assets:
- Instructors
- Administrative staff
- Mentors
- Cadre
- Facilities
- Funding

Activities

Pre-ChalleNGe:
- Administer orientation, drug testing, physicals, and placement tests
- Organize team-building training
- Counsel cadets and instruct on program expectations, life skills, and well-being

Residential phase:
- Coordinate cadet activities and fitness training
- Provide housing and meals
- Academic instruction
- Standardized testing, TABE/GED/HiSET
- Enforce appropriate cadet behavior and protocol
- Mentorship, form P-RAP
- Job skills instruction
- Exposure to vocations
- Drug testing and instruction on life skills and well-being (nutrition, hygiene, sexual health, substance abuse)
- Community service activities
- Track cadet progress
- Address parental concerns
- Graduate students
- Award credentials
- Register to vote and for Selective Service

Outputs

Cadet instruction:
- Cadets participate in activities and physical training
- Cadets housed, fed, and supervised
- Cadets instructed in classroom and learn independently
- Knowledge gained
- Cadets mentored
- Cadets meet behavior standards
- Cadets participate in job training
- Cadets tested for drugs and instructed in life skills and health
- Community service performed
- Increased awareness and perceived desirability of military service
- Cadets registered to vote and for Selected Service

Cadet graduated:
- Parental concerns addressed
- Cadet progress tracked
- Tests administered
- Cadets graduated
- Credentials or credit recovery awarded

Outcomes

Short term (0–3 years)

Cadets:
- Job and apprenticeship placements
- Postsecondary acceptance and attendance
- Military enlistment
- Improved health outcomes such as weight loss, smoking cessation, and physical fitness
- Life-coping skills such as leadership and self-discipline developed
- Cadets vote

Communities:
- Regular pools of reliable employees generated
- Increase in individuals participating in community service activities

Government and military:
- Increase in voter turnout
- Increase in high-quality enlistees

Medium term (3–7 years)

Cadets:
- College degree awarded
- Better cadet job skills and prospects
- Cadet career development
- Service to local communities
- Physical well-being

Communities:
- Employed, responsible individuals to support families
- Communities improved through community service
- Reduced unemployment
- Families and individuals who value civic participation
- Reduced drug addiction and crime

Government and military:
- Increase in skilled workforce
- Increase in civic engagement
- Higher regard for armed services passed on to peers and communities

Long term (7+ years)

Cadets:
- Increased civic participation
- Healthy social functioning and social interaction
- Economic self-sufficiency
- Physical well-being

Communities:
- Decreased rate of criminality
- Reduction in economic losses due to drug addiction
- More-livable communities
- Values passed on to peers, families, and communities

Government and military:
- Increased tax revenue
- Decreased expenditure on social services
- Increased appeal to corporations
- Greater involvement in government processes
- Increased enlistment from underrepresented populations

External factors and moderating factors: Parents, unexpected family events, job market, outside peer influence, cadet motivations, preexisting academic levels, prior criminality or drug use, preexisting mental or physical conditions

SOURCE: RAND analyses based on information collected from ChalleNGe sites. Figure also appears in Wenger et al., 2017.
NOTE: The Donohue intervention model was the initial design and description of the ChalleNGe program (Price, 2010). GED and HiSET credentials are awarded based on performance on standardized tests. The Post-Residential Action Plan is designed to support planning and goal development among cadets.

lesser extent, short-term outcomes (the right-hand side of the logic model). Ultimately, measuring the extent to which the program is meeting its mission will require collecting longer-term outcomes, similar to the type collected as part of the RCT described in Chapter One of this report (Bloom, Gardenhire-Crooks, and Mandsager, 2009; Millenky, Bloom, and Dillon, 2010; Millenky et al., 2011). Because of the expense and significant burden on participants, we do not intend to replicate an RCT. As we work to develop longer-term metrics, we will also explore methods of estimating the overall returns to the program that do not entail enrolling thousands of ChalleNGe cadets into RCTs. We discuss some of these efforts below.

Site Visits

Over the course of this project, the RAND team will visit each of the 39 current ChalleNGe sites. We developed a detailed protocol that includes questions for the director, deputy director, commandant (head of the cadre), recruiter, placement coordinator, mentoring coordinator, lead instructor (or principal), and the management information specialist. The protocol includes questions about the site's core mission, resources, staffing and hiring, outreach efforts, and relationship with the community, as well as many questions about the day-to-day operations and the types of data that are collected on cadets both during and after the program. The protocol also includes questions about recruiting and training mentors, cadet placements, the process of recruiting cadets, instruction, credentials, and occupational training opportunities. The protocol includes additional questions for sites in states with multiple ChalleNGe programs and for sites that offer Jobs ChalleNGe (a program with a focus on job training that is designed to follow ChalleNGe, currently available at the Michigan, South Carolina, and Georgia-Fort Stewart ChalleNGe sites). Table 3.1 shows the timing of our visits.

To date, we have visited 30 sites; we plan to visit the remaining 9 sites in 2019.[1] These site visits have served multiple purposes. During the first months of the project, we visited several sites to gather information as we developed our logic model. Site visits in the interim have been helpful as we worked to refine our research efforts in support of developing longer-term metrics of program effectiveness. From our site visits, we have learned quite a bit about the sources of variation across ChalleNGe sites, which has been helpful in designing our annual data collection instruments. We have also learned about the data maintained at each site, and we have been able to determine which sites are good candidates for future short-term pilot projects that would enable us to explore aspects of the ChalleNGe model that pose particular challenges, or pilot projects that could inform policies and practices across all sites. For example, at all site

---

[1]   We visited the Louisiana-Gillis Long site and the West Virginia site in late 2016, when we were designing our logic model. During the next year, we developed and considerably expanded our protocol; therefore, we returned to both West Virginia and Louisiana-Gillis Long in 2018.

Table 3.1
Schedule of Site Visits

ChalleNGe Site	2016–2017	2018	2019
Alaska		✓	
Arkansas			✓
California-Discovery	✓		
California-Grizzly	✓		
California-Sunburst		✓	
District of Columbia			✓
Florida	✓		
Georgia-Fort Gordon		✓	
Georgia-Fort Stewart		✓	
Georgia-Milledgeville		✓	
Hawaii-Barber's Point	✓		
Hawaii-Hilo	✓		
Idaho			✓
Illinois			✓
Indiana	✓		
Kentucky-Fort Knox (Bluegrass)		✓	
Kentucky-Harlan (Appalachian)	✓		
Louisiana-Camp Beauregard		✓	
Louisiana-Camp Minden		✓	
Louisiana-Gillis Long	✓	✓	
Maryland			✓
Michigan		✓	
Mississippi			✓
Montana	✓		
North Carolina-New London		✓	
North Carolina-Salemburg (Tarheel)		✓	
New Jersey	✓		
New Mexico	✓		
Oklahoma			✓

Table 3.1—Continued

ChalleNGe Site	2016–2017	2018	2019
Oregon	✓		
Puerto Rico		✓	
South Carolina		✓	
Tennessee			✓
Texas-East			✓
Texas-West[a]			✓
Virginia	✓		
Washington	✓		
Wisconsin	✓		
West Virginia	✓		
Wyoming	✓		

[a] The Texas ChalleNGe programs merged in late 2018; we will visit the single Texas program in 2019.

visits to date, program staff have reported varying degrees of difficulty when trying to gather accurate and timely information about graduate placement in the Post-Residential Phase. Sites typically rely on mentors to report this information, which is based on their regular interaction with graduates as their mentees. We are currently working with a site to develop other options for collecting placement data on graduates, such as through social media and alumni events.

We are also working with another site to improve mentor retention and engagement with mentees in the Post-Residential Phase. This effort is in response to a commonly expressed frustration across sites at the drop-off in mentor-mentee engagement in the three- to six-month period after graduation. We are currently working with the site to identify the mechanisms underlying the drop-off in engagement and ways to address this issue through mentor training and supports.

Other Analytic Efforts

Along with the annual reports to Congress, the development of the TOC and logic models, and the site visits described earlier, we are undertaking a series of related research efforts. Although these efforts jointly support the development of longer-term metrics and the goal of measuring ChalleNGe's effectiveness, each of these efforts is also focused on a specific topic or issue identified through site visits and engagement with program site leadership and staff. In the next section, we describe briefly our analytic efforts to date, focusing on two of the efforts that were undertaken in 2018: developing benchmarks for at-risk youth and examining the mentoring component of ChalleNGe. We also provide basic information, based on data collected in the annual survey, on CTE offerings. These data will be described in more detail in a separate report as part of the CTE analytic effort we are currently undertaking. We then briefly describe the remaining efforts and note other efforts that will likely be developed in the near future, depending on the needs of the ChalleNGe program.

Developing Benchmarks for At-Risk Youth

Developing longer-term metrics of cadet success is a primary focus of the overall project. To this end, we developed benchmarks to make longer-term metrics of cadet success more meaningful. Our specific goal was to estimate the likely outcomes for young people who do *not* complete the ChalleNGe program. Comparing the longer-term metrics of cadet success with these benchmarks will allow decisionmakers to better understand the effect of the ChalleNGe program, as well as the extent to which the program is meeting its overall mission. Of course, developing meaningful benchmarks requires analytic techniques designed to create appropriate comparison groups for ChalleNGe cadets.

We are currently finalizing a report intended to give ChalleNGe program site directors context for cadet family and schooling background and longer-term cadet outcomes, hereafter referred to as the *benchmark report*. In this effort, we use three nationally representative surveys of young adults and compare three populations: high school dropouts who do not attain a GED, high school dropouts who attain a GED, and high school graduates who do not attend college. We refer to these mutually exclusive groups as the *benchmark groups*. The report carefully presents each survey, and each benchmark group in each survey, and compares them across an array of metrics during two periods, the dropout window (ages 14–18) and the outcome window (ages 19–29).

The dropout window compares the benchmark groups at ages 14–18 across demographics, family background, behavioral history, scholastic achievement, and aptitude. It includes numerous tables that allow practitioners to compare their program population with the average dropout, GED holder, or traditional high school graduate in the United States. The outcome window compares the benchmark groups at ages 19–29 across education, employment, civic participation, health, and other outcomes. It also includes numerous tables that help practitioners form expectations of what constitutes a reasonable measure of success. Critically, the report then combines the data from all the surveys and all the groups to provide a range of expected outcomes at each age. Put another way, we take the means for each of the three benchmark groups (high school dropouts, GED holders, and high school graduates who do not attend college) and present the range of the lowest and highest sample mean. The range approximates for reasonable high-end and low-end expected outcomes for a high school dropout intervention program.

ChalleNGe cadets, even if they draw from the dropout population, are likely different from the average dropout: not only did they choose to participate in an intervention, but also they may have benefited from that intervention. Similarly, ChalleNGe cadets, even if successful in earning a credential or finishing high school, are likely different from the average graduate, given that they dropped out of school in the first place. Hence, we approximate low-end and high-end expected outcomes by comparing these groups. Often, the high and low estimates are set by dropouts and graduates, with GED holders falling in between, but this is not always the case. In addition, the report is careful to show rates or measures at each age, rather than an average over numerous ages. This precision is for two reasons. First, our aim in the benchmark analysis is to give a sense of expectations *over time*, or the progression of outcomes from ages 19–29 (a period marked by significant change, as our analysis will show). Second, our aim is also to give a sense of expectation at a *specific time*, so that a program can make specific age comparisons if it conducts a follow-up study.

For example, we show in Table 3.2 the range of arrest rates of the three benchmark groups, as self-reported in the National Longitudinal Survey of Youth (NLSY), from ages 19–29. This data does not cover the share who are charged or convicted, but rather the share who reported that they had been arrested at some point in the past year.

At a minimum, the youth population we study can expect to be arrested at a 5 percent rate throughout their 20s: The rate is highest (18 percent) for 21- and 22-year-olds, but it tapers off to 10 percent by age 29. The way that we interpret the first column in this table is that the expected arrest rate of a high school dropout at age 19 is 12 percent. An intervention, such as ChalleNGe, is successful if it reduces rates to a level below 12 percent. High school graduates are arrested at a 4 percent rate, which suggests that expecting an arrest rate for intervention program participants, such as cadets, to be below 4 percent is expecting cadets to be arrested at rates that are lower than the overall arrest rate of high school graduates. This likely represents an unreasonable expectation for programs and participants.

In Table 3.3, we show the range of nonemployment rates of the benchmark groups, as self-reported in the NLSY from ages 19–29. Here, *nonemployment* is defined as not having a job in the calendar year prior to the survey. As we discuss in the benchmarking report, the share not working in a year is very similar to the share who do not look for work in a year or are not attached to the labor force. Nonemployment is separate from unemployment, which requires an individual to be actively searching for formal work. There are many reasons why individuals choose not to participate in the workforce, but we would expect labor force participation to be high at young ages because it is a population in which disability rates are low, parenting rates and the need

Table 3.2
Self-Reported Arrest Rates Across Benchmark Groups, by Age (NLSY)

	19	20	21	22	23	24	25	26	27	28	29
Min arrest rate	4%	6%	6%	8%	6%	6%	5%	5%	5%	6%	5%
Max arrest rate	12%	14%	18%	16%	15%	14%	13%	13%	11%	10%	10%

NOTE: Data are from the 1997–2015 waves of the NLSY; "Min" and "Max" refer to the lowest and highest arrest rates at each age for three mutually exclusive groups: high school dropouts, high school dropouts who attain a GED, and high school graduates who do not attend college.

for dependent care coverage are low, and the groups we examine have extremely low rates of postsecondary enrollment, either by definition or in practice.

At a minimum, around one in six individuals in the youth population we study can expect to be not working throughout their 20s, but the expectation (or the average outcomes) is that a much higher share, around 35 percent on average, will not work at all in a year. As with Table 3.2, the way we interpret the first column of Table 3.3 is that the expected nonemployment rate of a high school dropout at age 19 is 39 percent. An intervention, such as ChalleNGe, is successful if it reduces rates to a level below 39 percent. However, the minimum for the benchmark groups was 20 percent. This suggests that expecting a nonemployment rate for intervention program participants, such as cadets, to be below 20 percent represents an unreasonable expectation for programs and participants.

The benchmarking report includes tables similar to Tables 3.2 and 3.3 for metrics of educational attainment and enrollment, employment, health, and civic participation. Our goal for this report is threefold: providing context for the background and outcomes of youth targeted by the ChalleNGe program, forming expectations of program effects, and providing something akin to a rule-of-thumb benchmark for programs to measure their own success.

Insights from the Literature and Site Visits on Mentorship

ChalleNGe uses a YIM approach, in which cadets are asked to nominate potential mentors from their social network. Mentors cannot be immediate family members of the cadet and must be at least 21 years old, of the same gender as the cadet, and live within a reasonable distance of the cadet. ChalleNGe performs a background check and delivers formal training to mentors prior to the official match. Once matched, cadets and their mentors are allowed to spend time off-site during the Residential Phase. Often, mentors and cadets perform community service together or meet to discuss future school and career options. Mentors play an important role in assisting cadets' reentry into the community after they graduate from the Residential Phase. Mentors and graduates are expected to sustain their relationship for 12 months, throughout the Post-Residential Phase. They are expected to be in contact on a weekly basis and have

Table 3.3
Share of Benchmark Groups Who Did Not Have a Job in the Prior Year, by Age (NLSY)

	19	20	21	22	23	24	25	26	27	28	29
Min not working	20%	17%	14%	15%	14%	15%	16%	18%	18%	20%	21%
Max not working	39%	35%	37%	34%	34%	31%	33%	33%	36%	38%	37%

NOTE: Data are from the 1997–2015 waves of the NLSY; "Min" and "Max" refer to the lowest and highest nonemployment rates at each age for three mutually exclusive groups: high school dropouts, high school dropouts who attain a GED, and high school graduates who do not attend college.

in-person meetings at least twice a month. Program staff monitor the matches during the Post-Residential Phase and provide support to mentors and mentees if necessary.

There is emerging evidence to support the use of YIM (Schwartz et al., 2013; Spencer et al., 2016). To date, the most rigorous evaluations of YIM have focused exclusively on the ChalleNGe program. Most ChalleNGe cadets were able to nominate an adult to serve as their mentor, which suggests the feasibility of YIM, and ChalleNGe participants who chose their own mentors had longer relationships with their mentors in comparison with those who had mentors assigned to them (Schwartz et al., 2013). Furthermore, youth who had longer mentoring relationships reported better outcomes in comparison with those who had shorter mentoring relationships (Schwartz et al., 2013). Finally, qualitative findings provide insights into the quality of the YIM relationships. Mentees reported feeling supported by their mentors, as well as a sense of closeness. Mentees also shared that their mentors had helped them make progress toward their post-ChalleNGe goals, become more confident, and be more capable of having positive relationships with adults (Spencer et al., 2016).

Although YIM has shown some promising results, there are a few limitations. Spencer et al. (2016) found that mentees reported limited access to resources that were outside of their social network, a finding that confirms the hypothesis that YIM is not best at increasing youth's social capital or connections that extend a youth's personal and professional opportunities. In addition, ongoing training and mentor support are critical to developing meaningful and effective YIM relationships, despite the fact that mentors and mentees already know each other prior to the start of their mentoring relationships. In the next section, we draw on findings about mentoring based on our site visits.

Insights on Mentoring from the Analysis of Site Visits

These findings incorporate analysis from 15 site visits conducted in 2017 and seven site visits conducted in 2018.[2] The purpose of these visits is to develop effective working

---

[2] Since the completion of the analysis, the team has conducted additional visits. The sites included are the Youth ChalleNGe Academies in Georgia, Kentucky, Louisiana, Hawaii, Washington, Wyoming, California, Montana, New Mexico, Wisconsin, Indiana, Florida, Virginia, and New Jersey.

relationships with sites and staff and to gather data about the operation and programming at each site. During each site visit, we conducted interviews with staff members from various departments, with an overarching goal of learning about different programming at the site and identifying resources and challenges. In the following section, we analyze site visit notes from these 22 sites. A preliminary coding scheme was developed based on existing literature on YIM and youth mentoring in general. The coding focused on mentor selection and recruitment, mentor training, mentor responsibilities, mentoring relationships, and mentor impact. These categories were chosen because existing research—on YIM specifically and youth mentoring more broadly—has found these factors to be important components of an effective mentoring program or critical to evaluating a mentoring program (or both). Line-by-line coding was used to extract information from the site visit notes. The coding scheme was revised as new codes emerge, and notes were reread according to the new coding structure; this process was repeated until no new codes were identified (Corbin and Strauss, 1990). Categories of similar codes were created and developed into themes. The final list of themes included recruitment of mentor, mentor selection criteria, matchmaking, mentor training, mentor responsibilities, mentoring relationships, mentor impact, and challenges and facilitators to implementation. There are substantial variations in how mentoring is implemented across the sites.

Recruitment of Mentor, Mentor Criteria, and Matchmaking

Because the ChalleNGe program uses the YIM model, most cadets arrived at their program site having identified a mentor. However, almost all sites had to assign mentors to a proportion of cadets, either because the cadets did not have someone identified or because the adults they nominated did not meet the criteria. The number of cadets assigned a mentor by staff ranged from a few cadets per class to close to 40 percent of a class. Given the need to assign mentors to some cadets, sites typically had a pool of potential mentors. Sites recruited potential mentors mainly through community outreach, such as by visiting high schools, nonprofit organizations, churches, and business groups.

Four sites expressed doubts about the feasibility of YIM. The main concern was that sometimes cadets could not name any adults to serve as a mentor. Staff from a few sites suggested that if the cadets had a mentor, they probably would not need to participate in ChalleNGe. However, most sites were very enthusiastic about YIM and said that "warm matches" (i.e., when a mentor was nominated by cadet or family) were generally more successful than "cold matches" (i.e., when a mentor was assigned by the program). These sites also reported that almost all cadets came to the program with a mentor, or that the sites required applicants to name a mentor on their application.

As mentioned above, mentors must pass a background check. Program staff at all sites noted that mentors should be at least 21 years old, and staff at six sites suggested that the best mentors were grandparents or other nonparental family members.

In almost all cases, mentors and cadets were matched by gender; staff at one site said that exceptions could be made for a male cadet to be matched with a female mentor. At another site, staff discussed the importance of cultural matching and noted that it was often difficult to find mentors who were from the same cultural group as the cadets.

We found variation across ChalleNGe sites in terms of what staff considered to be the most important determinants of a good mentor. At two sites, staff reported that adults who were in a helping profession, such as a police officer or social worker, and/or who worked with youth, such as teachers, coaches, and counselors, often were the best mentors. However, staff at one site reported that adults who worked with youth were generally too busy and had little time to commit to a mentee. Staff at six sites said that relatives (e.g., grandparents, uncles) and family friends worked well as mentors because they already had a relationship with the cadet and were invested in the success of the cadet. At one site, staff said that adults who had experienced adversity and bounced back were the best mentors because cadets could relate to them. At another site, staff suggested that younger mentors were more relatable to cadets.

Mentor Training and Mentor Responsibilities

All sites provided training to mentors. The most typical training is a one-day, on-site training. Two sites provided a three-day training. Mentors would travel to the ChalleNGe site in the morning. In the afternoon, they would have time to meet with their mentees, and some sites would have joint training for mentors and mentees. Four sites mailed training materials to mentors if they were not available to attend in person. The timing of the trainings varied. Training started as early as week five or six and as late as week nine or ten of the Residential Phase. All interviewees reported that in-person training was more effective because mentors were able to experience the residential component of ChalleNGe and meet with cadets face-to-face to start developing a mentoring relationship.

Training varied across sites. At one site, training included skill-based exercises, such as active listening. At another site, staff discussed mentoring best practices, communication skills, and setting expectations with mentors during their training. Finally, staff at one site noted that they recently added case management training and thought the training was helpful to mentors. At four sites, staff thought their current training was not sufficient. Staff at one site would like to train their mentors on trauma, behavioral issues (e.g., gang involvement, substance use), and how to work with young people. No staff discussed ongoing or follow-up training with mentors, including during the Post-Residential Phase.

Mentors were asked to commit to at least one year postgraduation. All sites required mentors to submit monthly reports. Two sites developed a web portal for mentors to upload the monthly report. Other sites would ask mentors to either email, mail, or text the monthly reports to case managers. In the monthly report, mentors were asked to report cadets' progress on achieving their post-residential goals as outlined in their P-RAP. Sometimes mentors were asked to submit cadets' paystubs or call cadets' employers to verify employment. In addition, mentors were asked to contact cadets once a week and were directed that two of the four monthly contacts should be in-person visits. However, staff reported that most contacts between cadets and mentors were via text messaging or social media (e.g., Facebook).

Mentoring Relationship

Eight sites hosted activities during the Residential Phase to foster positive relationships between cadets and mentors. Two sites described an on-site matching ceremony in which mentors and cadets were matched and spent time getting to know each other for a day. After mentors had completed required background checks and training, many sites invited them to visit cadets on-site. Site visits were as frequent as once a week. All of the sites that organized activities for mentors and cadets during the Residential Phase required cadets to either write or call mentors weekly or biweekly. Three sites allowed mentors to take cadets off-site. Two sites sent monthly newsletters to mentors during the Residential Phase to inform them about cadets' progress and activities.

At many sites, staff said that developing a close and strong mentoring relationship during the Post-Residential Phase was a challenge. The most commonly cited issues were lack of commitment from mentors, disappointment from mentors when graduated cadets were not progressing, indifference from cadets, and instability among graduated cadets (e.g., cadets would move or change their phone number). Moreover, mentors were overwhelmed by the responsibilities. Mentors did not have time to keep up with the monthly reports and weekly contacts. As mentioned previously, incomplete monthly reporting was the most frequently cited challenge by sites. To facilitate communication and relationship building between mentors and graduated cadets during the Post-Residential Phase, two sites organized activities, such as a career fair or a pizza night, to encourage mentors and graduated cadets to meet and stay in touch.

Mentor Impact

Staff at four sites discussed mentor impact on cadets. At two sites, staff reported that the mentors' relationships with cadets were the reason why some of them stayed during the Residential Phase. Mentors were among the few people from home that cadets kept in contact with while they were in the program. Staff at one site talked about their peer mentor program and the perceived impact of the program: Peer mentors were cadets from previous classes and guided new cadets through their initial stage. Staff at this site highly recommended the peer mentor approach and said that peer mentors were their strongest "retention tool."

Challenges and Facilitators to Implementation

One of the most-cited challenges was retention of mentors. At four sites, staff reported that mentors' lack of commitment to the program and to the cadet was a barrier to retention. The problem varied from losing 20 percent of mentors to almost 80 percent during the Post-Residential Phase. One staff member described the problem this way: "The [mentoring] program doesn't work the way it should. Mentors are not given extensive training and are often unmotivated . . . . If the cadet no longer wants contact, then the mentor doesn't push it. It matters if the mentor has a connection with the cadet."

Staff at almost all sites reported that the extensive reporting requirements were barriers to retention. Program staff have tried to make the reporting as easy as possible. Staff at three sites suggested simplifying the monthly reporting forms to ask fewer questions. In addition to having difficulty engaging mentors, staff at two sites talked about the challenge of getting buy-in from parents. Parental buy-in was particularly important to the recruitment process because programs often relied on parents to nominate mentors for their child if the cadet did not identify a mentor.

Staff also discussed strategies that they had used to address some of the aforementioned challenges. For example, staff at three sites said that they had implemented mentor appreciation activities, such as a "mentor of the year award" and a Facebook mentor page, to increase mentor engagement. Moreover, staff noticed that mentor engagement decreased substantially postgraduation, and some addressed this issue through organized, structured activities for mentors and mentees during the Post-Residential Phase. Also, staff at three sites discussed the possibility of using technology to improve mentor and mentee communication. Given the responsibilities asked of mentors, staff emphasized the need to clearly communicate expectations to mentors so that there would be "no surprises." Staff at one site conducted exit interviews with mentors at the end of the Post-Residential Phase to gather feedback. Finally, staff at two sites were in favor of the idea of providing incentives to mentors to promote commitment and retention.

Mentoring has the potential to make a significant contribution to the ChalleNGe program and to promote positive outcomes in cadets. The YIM approach applied by ChalleNGe has shown promise in terms of sustaining the positive impact of ChalleNGe. However, there are barriers to implementing a robust and sustainable mentoring program, as indicated by findings from site visits. Additional mentor training and further development of the program during the Post-Residential Phase could be particularly helpful. Moreover, systematic evaluation is needed to measure the effectiveness of the mentoring program.

Mentoring is a key feature of the ChalleNGe program and it supports the program's eight core components. A review of the literature suggests that mentoring can have an important impact on young people's lives. YIM, the process by which most cadets in ChalleNGe are matched with a mentor, is generally associated with a longer-lasting match, especially if it is based on an existing relationship between the mentor and mentee. However, findings from our site visits also pointed to the weaknesses and areas in need of improvement in the mentoring program, especially as it relates to the longer-term viability of the mentor-mentee relationship and mentor reporting on post-residential placement of cadets.

Pilot Programs

In addition to the activities described above, the RAND research team is partnering with ChalleNGe sites to initiate pilot programs to inform improvements to ChalleNGe programming in the areas of mentoring and post-residential placement tracking. The overall goal of the pilots is to identify improvements that all ChalleNGe programs can benefit from.

Mentoring pilot

The RAND research team is currently working with a ChalleNGe site to develop a mentoring pilot and qualitatively assess its effects. Our discussions with staff at the pilot site about the design of the mentoring pilot are ongoing. Currently, the main objective of the pilot would be to strengthen the initial and ongoing training that mentors receive to better prepare and support them in their role. The ChalleNGe site will be matched with publicly accessible mentor training resources that can be used to tailor the training and supports that mentors receive. These changes will be qualitatively assessed by the program with support from the RAND team.

ChalleNGe Alumni Survey Pilot

The RAND research team is working with another site to identify means of following and gathering data on cadets postgraduation using a simple survey to collect placement data and information on longer-term outcomes, such as education, employment, and life transitions. Our site visits suggest that all ChalleNGe sites face difficulties in following and gathering data on cadets in the Post-Residential Phase, let alone beyond that phase. Much of the information gathered on cadet well-being is anecdotal and not representative of the range of experiences graduates are likely to have. We will work with the site to develop and test a data collection instrument that can efficiently and effectively collect information on graduates and generate findings that are more broadly representative of postgraduation education, employment, and life experiences.

Examining Job Training and Career and Technical Education

As noted earlier, the ChalleNGe program includes eight core components, one of which is job skills. This component is also emphasized across a wide variety of programs aimed at disadvantaged youth. CTE, historically referred to as vocational education, has been receiving more national attention over the past decade as evidenced by the passage of the Carl D. Perkins Career and Technical Education Act of 2006 and the Workforce Innovation and Opportunity Act of 2014, and CTE was among the topics we discussed in each of our site visits.[3] The RAND research team is examining the relevant literature with the goal of identifying practices that have been found effective in other programs. An in-depth review of this literature will allow us to determine

---

[3]   The U.S. Department of Education's Institute of Education Sciences notes this surge in interest related to research on CTE as well (Larson, 2018).

the extent to which the ChalleNGe sites follow preferred practices, and it also may suggest ways to strengthen individual ChalleNGe sites or the ChalleNGe program as a whole. We also collected data from all the sites in our most recent data collection on CTE course and certification opportunities. In Table 3.4, we provide information on the types of courses provided and whether certification is offered.

A relatively small share of cadets participate in CTE. According to the data we collected, approximately 11 percent of cadets who entered ChalleNGe earned some kind of CTE credit. Staff reported that 14 percent of cadets who entered ChalleNGe earned a certificate or credential. More cadets earn some kind of a credential or certificate that does not include earning formal CTE credit, which could be counted as credit toward a high school diploma. The most commonly provided vocational courses have to do with food safety, construction, workplace safety, production and welding, auto mechanics, forklift operations, health care, and other fields (Table 3.4).[4]

We will continue to analyze the data collected, hold CTE-focused meetings with site stakeholders to better understand CTE programming, and incorporate lessons learned from the literature. This effort should be completed by mid-2019.

Relationship Between Program Characteristics and Program Outcomes

This analysis will use the data that form the basis of this series of reports (the data collected each year from each ChalleNGe program) to estimate the relationships between program characteristics and short- and longer-term outcomes. This effort will also include the development of a value-added–type model to examine the characteristics that are associated with higher TABE score gains. These efforts should be completed by mid-2019.

Summary

In this chapter, we reviewed our logic model. We also described the schedule of ChalleNGe site visits, including the methodology that we are using to collect and analyze data from the sites. Finally, we described our additional analytic efforts, all of which support the operation of the ChalleNGe program and will assist in determining the extent to which the program is meeting its mission. Taken as a group, these analytic efforts will examine many of the core components of ChalleNGe and will produce actionable recommendations to improve program outcomes. In Chapter Four, we offer some concluding thoughts.

---

[4]   The question on vocational courses provided was posed to all sites in reference to training provided during ChalleNGe. At the time of this publication, several sites participating in the survey provide Job ChalleNGe (Michigan, Georgia, South Carolina) or other vocational training programs after program completion (as in the case of Alaska).

Table 3.4
Site-Reported Sample of Career and Technical Education and Training Offerings

CTE Field	Sites Offering Courses	Examples of Certifications Offered	Sites Offering a Certification
Food preparation	9	ServSafe	7
Construction	7	NCCER, basic carpentry, and fall protection	5
Workplace safety	6	OSHA10	5
Computer and information technology, including web design	6	Microsoft Office, 3D ThinkLink, and computer training	3
Industrial manufacturing and welding	5	Basic welding certification	2
Visual arts and media	5	Production tech, visual arts	2
Auto mechanics and automotive body repair	4	Basic auto mechanics	1
Forklift operation	3	Forklift operator certification	3
Cosmetology and barbering	3	Beginning cosmetology	1
Small engine repair	2	National small engine repair certification	2
Cardiopulmonary resuscitation (CPR)	2	CPR	2
Phlebotomy	2	Phlebotomy	2
Certified nursing assistant (CNA)	2	CNA	2
Heavy equipment operator	2	Heavy equipment operator	2
Masonry	2	Masonry and industrial maintenance	1
Economic literacy and personal finance	2	Economic literacy	1
Clerical and office	2	Clerical and office	1

NOTES: This table is based on data reported by the sites. It is possible, for example, that some sites offer food preparation courses that award a ServSafe certification but did not report this information.

CHAPTER FOUR

Concluding Thoughts

The National Guard Youth ChalleNGe program provides opportunities for young people who struggle in traditional high schools. Cadets who participate in ChalleNGe have the opportunity to earn various academic credentials: certificates (such as the GED or HiSET), high school credits that allow them to reenroll in their home high schools, or high school diplomas. Along with academic credentials, cadets can gain a variety of occupational experience and training, perform community service, improve their physical fitness, and develop their noncognitive or socioemotional skills. The ChalleNGe model, with its focus on eight core components, is designed to increase the likelihood that cadets gain a variety of experiences and skills that prepare them to get their lives back on track. Previous studies have shown that ChalleNGe is effective (Bloom, Gardenhire-Crooks, and Mandsager, 2009; Millenky, Bloom, and Dillon, 2010; Millenky et al., 2011) but continuous monitoring is needed to identify areas in need of improvement.

To date, more than 220,000 young people have participated in ChalleNGe, and about 165,000 have completed the program. ChalleNGe programs took in more participants in 2017 than in 2016. Some of this increase is due to new programs providing data—e.g., in California, Georgia, and Tennessee—but some is also due to existing programs enrolling more cadets. Among cadets who entered ChalleNGe in 2017, nearly 10,000 graduated from one of the 39 sites included in our data collection, and nearly 4,000 received at least one education credential (and this number is over 6,200 if high school credits are counted as well). Data collected from the sites demonstrate that ChalleNGe cadets show marked improvement in academic skills (measured by the proportion who achieve key benchmarks on the TABE) and in the program's other core components. The overall graduation rate, however, remained roughly constant, as did the proportion of graduates who scored at or above the ninth-grade level on the standardized test used to track cadet progress. While more young people are graduating from ChalleNGe overall, academic quality appears to be roughly constant over the period.

In the first two years of the RAND team's current ChalleNGe project, we have collected three rounds of data from the ChalleNGe sites, and we have carried out 30 site visits. This report is the third in support of ChalleNGe's annual reporting requirement to Congress. We will produce one more annual report in 2020. We are also making progress on the second goal of the project: developing metrics to measure cadets' outcomes. These metrics will help to inform assessments of how well the ChalleNGe program is doing in meeting its stated mission "to intervene in and reclaim the lives of 16–18-year-old high school dropouts, producing program graduates with the values, life skills, education, and self-discipline necessary to succeed as productive citizens" (National Guard Youth ChalleNGe, 2010. p. 7).

In addition to collecting annual program- and cadet-level indicators, the RAND research team is carrying out several other analytic efforts, as discussed in this report. These efforts include research on a variety of topics that are relevant to ChalleNGe's core components, including mentorship, benchmarking, and CTE (also referred to as vocational education and training). The research team will also continue to work on developing program-level indicators and exploring associations with outcomes of interest, including performance on TABE, graduation, and post-residential placement. We shared early findings from two of the analytic efforts: benchmarking and mentoring.

The first analytic effort focuses on developing benchmarks to facilitate comparison between youth population in a specific program with the average dropout, GED holder, or traditional high school graduate in the United States. Because this effort does not involve new data collection, developing and examining benchmarks is a relatively inexpensive way to track progress of cadet graduates on key outcomes of interest. The results of this study can help form expectations of program effects and provide a rule-of-thumb benchmark for programs to measure their own success.

The second analytic effort focuses on examining the literature on mentoring and sharing insights on the implementation of mentoring from our site visits, with a focus on YIM. The review of literature suggests that YIM is more likely to lead to long-term and more-resilient mentor-mentee relationships, particularly if they are built on an existing relationship. Despite the overall positive assessment of YIM, some limitations were also highlighted. Most notably, mentees reported limited access to resources that were outside of their social network if their mentor was identified from within their own network. Mentoring is also challenged overall because of limited training and supports provided to mentors in the Post-Residential Phase, competing obligations, and general lack of interest in sustaining the relationship on the part of the mentee, mentor, or both.

In addition to these analytic efforts, the research team is also working with two sites to develop and field pilots. One is focused on developing and tracking new approaches to implementing the mentoring program, and the other is focused on designing and administering an alumni survey to supplement information collected on post-residential placement. Both will include an evaluation component, the results of which are intended to help ChalleNGe programs track their progress and inform implementation of program improvements.

Site-Specific Information

This appendix includes a complete list of the ChalleNGe programs, as well as the program-level tables of information. Table A.1 provides the complete name and location (state) of each program.

Tables A.2–A.41 include detailed information collected from each program. We carried out data collection in September–October 2018. The focus of the data collection was on classes that began in 2017 (Classes 48 and 49 according to the ChalleNGe class numbering system, which began with the first class in the 1990s). Data were collected from three additional ChalleNGe programs (California-Discovery, Georgia-Milledgeville, and Tennessee) that had only recently begun during the last data collection in 2017 and therefore did not report information for the 2017 report. Similarly, we received data in 2018 from the Puerto Rico ChalleNGe program, which had been negatively affected by Hurricane Maria in September 2017 and did not report data last year.

In some cases, programs provided incomplete data or data that were suspect in some way. For example, counts reported at the program level do not match aggregate counts at the cadet level. When a site provided incomplete data, we indicated the specific elements that were not reported. Some of these data issues are related to the variation in how the individual sites collect and store data. RAND analysts are currently exploring strategies to increase the accuracy of future data collected from the sites, with a focus on limiting the burden of data collection on sites and ChalleNGe personnel.

The sites are listed alphabetically by state or territory name. Each table includes metrics of the number and type of staff, total funding in 2017, and the numbers of cadets who applied, entered, graduated, and received various credentials. The tables also include data related to several of the core components—service to community (and calculated values based on local labor market conditions), gains on specific physical fitness tests, and the numbers of cadets registered to vote and registered for Selective Service. Finally, the tables include information about postgraduation placement (although there is no information on Class 49's 12-month placement rates because fewer than 12 months had passed since graduation for this group). The tables also include 12-month placement rates for Class 47; at the time of our previous data collection, 12-month information was not yet available for cadets in Class 47.

Table A.1
National Guard Youth ChalleNGe Program Abbreviation, State, Name, and Type

Program Abbreviation	State	Program Name	Program Type[a]
AK	Alaska	Alaska Military Youth Academy	GED
AR	Arkansas	Arkansas Youth ChalleNGe	GED
CA-DC	California	Discovery ChalleNGe Academy	High school credits or diploma
CA-LA	California	Sunburst Youth Academy	High school credits or diploma
CA-SL	California	Grizzly Youth Academy	High school credits or diploma
D.C.	District of Columbia	Capital Guardian Youth ChalleNGe Academy	High school credits or diploma
FL	Florida	Florida Youth ChalleNGe Academy	GED
GA-FG	Georgia	Fort Gordon Youth ChalleNGe Academy	GED
GA-FS	Georgia	Fort Stewart Youth ChalleNGe Academy	GED
GA-MV	Georgia	Milledgeville Youth ChalleNGe Academy	GED
HI-BP	Hawaii	Hawaii Youth ChalleNGe Academy at Barber's Point	GED
HI-HI	Hawaii	Hawaii Youth ChalleNGe Academy at Hilo	High school credits or diploma
ID	Idaho	Idaho Youth ChalleNGe Academy	High school credits or diploma
IL	Illinois	Lincoln's ChalleNGe Academy	GED
IN	Indiana	Hoosier Youth ChalleNGe Academy	GED
KY-FK	Kentucky	Bluegrass ChalleNGe Academy	High school credits or diploma
KY-HN	Kentucky	Appalachian ChalleNGe Program	High school credits or diploma
LA-CB	Louisiana	Louisiana Youth ChalleNGe Program—Camp Beauregard	GED
LA-CM	Louisiana	Louisiana Youth ChalleNGe Program—Camp Minden	GED
LA-GL	Louisiana	Louisiana Youth ChalleNGe Program—Gillis Long	GED
MD	Maryland	Freestate ChalleNGe Academy	High school credits or diploma
MI	Michigan	Michigan Youth ChalleNGe Academy	High school credits or diploma

Table A.1—Continued

Program Abbreviation	State	Program Name	Program Type[a]
MS	Mississippi	Mississippi Youth ChalleNGe Academy	High school credits or diploma
MT	Montana	Montana Youth ChalleNGe Academy	GED
NC-NL	North Carolina	Tarheel ChalleNGe Academy—New London	GED
NC-S	North Carolina	Tarheel ChalleNGe Academy—Salemburg	GED
NJ	New Jersey	New Jersey Youth ChalleNGe Academy	High school credits or diploma
NM	New Mexico	New Mexico Youth ChalleNGe Academy	GED
OK	Oklahoma	Thunderbird Youth Academy	High school credits or diploma
OR	Oregon	Oregon Youth ChalleNGe Program	High school credits or diploma
PR	Puerto Rico	Puerto Rico Youth ChalleNGe Academy	High school credits or diploma
SC	South Carolina	South Carolina Youth ChalleNGe Academy	GED
TN	Tennessee	Volunteer Youth ChalleNGe Academy	GED
TX-E	Texas	Texas ChalleNGe Academy—East	High school credits or diploma
TX-W	Texas	Texas ChalleNGe Academy—West	High school credits or diploma
VA	Virginia	Virginia Commonwealth ChalleNGe Youth Academy	GED
WA	Washington	Washington Youth Academy	High school credits or diploma
WI	Wisconsin	Wisconsin ChalleNGe Academy	High school credits or diploma
WV	West Virginia	Mountaineer ChalleNGe Academy	High school credits or diploma
WY	Wyoming	Wyoming Cowboy ChalleNGe Academy	GED

NOTE: In late 2018, the two Texas ChalleNGe sites consolidated into a single site.

[a] Program type is based on the most common credential reported by the sites.

Some of the data in the following tables (along with other cadet-level data collected at the same time) formed the basis of analyses presented in Chapter Two. These same data will also be used in some of our future analyses described in Chapter Three.

Table A.2
Profile of Alaska Military Youth Academy

ALASKA MILITARY YOUTH ACADEMY, ESTABLISHED 1994					
Graduates Since Inception: 5,457			**Program Type: Credit Recovery, High School Diploma, GED**		

Staffing

	Instructional	Cadre	Administrative	Case Managers	Recruiters	Total
Number employed	8	25	19	5	3	60

Funding

	Federal Funding	State Funding	Other Funding
Classes 48 and 49	$3,715,000	$1,238,333	$3,826,400

Residential Performance

	Dates	Applied	Entered Pre-ChalleNGe	Graduated	Received GED/HiSET	Received HS Credits	Received HS Diploma
Class 48	March 2017–July 2017	394	220	180	102	0	6
Class 49	Sept. 2017–Jan. 2018	331	188	146	68	0	3

Physical Fitness

	Push-Ups		1-Mile Run		BMI	
	Initial	Final	Initial	Final	Initial	Final
Class 48	31.5	42.9	09:54	08:04	26.0	27.6
Class 49	22.9	39.2	09:58	08:17	25.7	*

Responsible Citizenship

	Voting		Selective Service	
	Eligible	Registered	Eligible	Registered
Class 48	27	27	21	21
Class 49	34	34	22	22

Table A.2—Continued

Service to Community			
	Service Hours per Cadet	**Dollar Value per Hour**	**Total Value**
Class 48	56	$27.45	$276,696
Class 49	51	$27.45	$204,393

Post-Residential Performance Status							
	Graduated	**Contacted**	**Placed**	**Education**	**Employment**	**Military**	**Multiple/ Other**
Class 47							
Month 12	165	81	79	25	22	1	31
Class 48							
Month 1	180	142	140	95	19	1	25
Month 6	180	113	112	70	15	1	26
Month 12	180	67	65	9	35	0	23
Class 49							
Month 1	146	107	106	78	10	2	16
Month 6	146	N/A	N/A	N/A	N/A	N/A	N/A

* Did not report; HS = high school; N/A = not available.

Table A.3
Profile of Arkansas Youth Challenge

ARKANSAS YOUTH CHALLENGE, ESTABLISHED 1993					
Graduates Since Inception: 3,747			Program Type: GED		

Staffing						
	Instructional	**Cadre**	**Administrative**	**Case Managers**	**Recruiters**	**Total**
Number employed	5	21	15	4	4	49

Funding			
	Federal Funding	**State Funding**	**Other Funding**
Classes 48 and 49	$2,450,250	$816,750	$0

Table A.3—Continued

Residential Performance

	Dates	Applied	Entered Pre-ChalleNGe	Graduated	Received GED/HiSET	Received HS Credits	Received HS Diploma
Class 48	Jan. 2017–June 2017	250	153	103	34	0	0
Class 49	July 2017–Dec. 2017	261	159	88	35	0	0

Physical Fitness

	Push-Ups		1-Mile Run		BMI	
	Initial	Final	Initial	Final	Initial	Final
Class 48	31.4	47.5	10:11	08:01	23.5	*
Class 49	18.2	36.9	11:31	10:50	24.4	*

Responsible Citizenship

	Voting		Selective Service	
	Eligible	Registered	Eligible	Registered
Class 48	20	17	41	0
Class 49	17	0	35	19

Service to Community

	Service Hours per Cadet	Dollar Value per Hour	Total Value
Class 48	81	$20.01	$167,092
Class 49	71	$20.01	$124,453

Post-Residential Performance Status

	Graduated	Contacted	Placed	Education	Employment	Military	Multiple/Other
Class 47							
Month 12	95	95	57	28	18	4	7
Class 48							
Month 1	103	94	46	23	16	0	7
Month 6	103	96	79	44	21	3	11
Month 12	103	89	77	29	27	7	14
Class 49							
Month 1	88	84	67	47	12	1	7
Month 6	88	81	67	32	18	1	16

* Did not report; HS = high school.

Table A.4
Profile of Discovery ChalleNGe Academy (California)

DISCOVERY CHALLENGE ACADEMY, ESTABLISHED 2017

Graduates Since Inception: 231	Program Type: Credit Recovery, High School Diploma

Staffing

	Instructional	Cadre	Administrative	Case Managers	Recruiters	Total
Number employed	8	23	14	3	1	49

Funding

	Federal Funding	State Funding	Other Funding
Classes 48 and 49	$4,025,294	$1,010,831	$3,451,830

Residential Performance

	Dates	Applied	Entered Pre-ChalleNGe	Graduated	Received GED/HiSET	Received HS Credits	Received HS Diploma
Class 48	Jan. 2017–June 2017	214	140	105	0	75	30
Class 49	July 2017–Dec. 2017	257	147	126	0	96	30

Physical Fitness

	Push-Ups		1-Mile Run		BMI	
	Initial	Final	Initial	Final	Initial	Final
Class 48	31.2	49.7	09:12	07:58	24.9	23.5
Class 49	26.1	34.4	10:03	08:34	26.0	24.8

Responsible Citizenship

	Voting		Selective Service	
	Eligible	Registered	Eligible	Registered
Class 48	12	12	12	12
Class 49	7	7	7	7

Service to Community

	Service Hours per Cadet	Dollar Value per Hour	Total Value
Class 48	56	$29.09	$171,049
Class 49	62	$29.09	$227,251

Table A.4—Continued

Post-Residential Performance Status

	Graduated	Contacted	Placed	Education	Employment	Military	Multiple/ Other
Class 47							
Month 12	N/A	N/A	N/A	N/A	N/A	N/A	N/A
Class 48							
Month 1	105	105	99	74	19	1	5
Month 6	105	105	102	62	23	5	12
Month 12	105	105	102	39	28	9	26
Class 49							
Month 1	126	126	123	88	17	0	18
Month 6	126	126	123	74	17	2	30

* Did not report; HS = high school; N/A = not available.

Table A.5
Profile of Sunburst Youth Academy (California)

SUNBURST YOUTH ACADEMY, ESTABLISHED 2008					
Graduates Since Inception: 3,202			**Program Type: Credit Recovery, High School Diploma, HiSET**		

Staffing

	Instructional	Cadre	Administrative	Case Managers	Recruiters	Total
Number employed	22	25	23	4	0	74

Funding

	Federal Funding	State Funding	Other Funding
Classes 48 and 49	$5,750,000	$1,916,667	$2,736,436

Residential Performance

	Dates	Applied	Entered Pre-ChalleNGe	Graduated	Received GED/HiSET	Received HS Credits	Received HS Diploma
Class 48	Jan. 2017–June 2017	326	206	172	0	155	17
Class 49	July 2017–Dec. 2017	316	225	193	0	175	18

Table A.5—Continued

Physical Fitness

	Push-Ups		1-Mile Run		BMI	
	Initial	Final	Initial	Final	Initial	Final
Class 48	20.2	51.6	09:23	07:34	27.5	26.9
Class 49	24.3	51.4	09:23	07:31	26.0	25.7

Responsible Citizenship

	Voting		Selective Service	
	Eligible	Registered	Eligible	Registered
Class 48	25	25	36	36
Class 49	41	41	42	42

Service to Community

	Service Hours per Cadet	Dollar Value per Hour	Total Value
Class 48	44	$29.09	$220,153
Class 49	52	$29.09	$291,947

Post-Residential Performance Status

	Graduated	Contacted	Placed	Education	Employment	Military	Multiple/Other
Class 47							
Month 12	197	197	179	101	26	6	46
Class 48							
Month 1	172	166	143	111	20	2	10
Month 6	172	166	159	117	29	6	7
Month 12	172	160	153	117	21	10	5
Class 49							
Month 1	193	143	134	120	14	0	0
Month 6	193	153	143	110	23	6	4

* Did not report; HS = high school.

Table A.6
Profile of Grizzly Youth Academy (California)

GRIZZLY YOUTH ACADEMY, ESTABLISHED 1998

| Graduates Since Inception: 5,865 | Program Type: Credit Recovery, High School Diploma, HiSET |

Staffing

	Instructional	Cadre	Administrative	Case Managers	Recruiters	Total
Number employed	12	28	12	4	2	58

Funding

	Federal Funding	State Funding	Other Funding
Classes 48 and 49	$5,625,000	$1,875,000	$822,000

Residential Performance

	Dates	Applied	Entered Pre-ChalleNGe	Graduated	Received GED/HiSET	Received HS Credits	Received HS Diploma
Class 48	Jan. 2017–June 2017	294	204	186	0	149	37
Class 49	July 2017–Dec. 2017	334	215	195	0	149	46

Physical Fitness

	Push-Ups		1-Mile Run		BMI	
	Initial	Final	Initial	Final	Initial	Final
Class 48	21.1	40.1	09:54	07:39	*	*
Class 49	25.7	40.6	08:44	07:09	*	*

Responsible Citizenship

	Voting		Selective Service	
	Eligible	Registered	Eligible	Registered
Class 48	30	30	19	19
Class 49	27	27	18	18

Service to Community

	Service Hours per Cadet	Dollar Value per Hour	Total Value
Class 48	52	$29.09	$280,276
Class 49	54	$29.09	$308,019

Table A.6—Continued

Post-Residential Performance Status

	Graduated	Contacted	Placed	Education	Employment	Military	Multiple/ Other
Class 47							
Month 12	203	203	161	63	55	12	31
Class 48							
Month 1	186	186	167	122	9	1	35
Month 6	186	185	161	82	16	2	61
Month 12	186	186	147	59	31	4	53
Class 49							
Month 1	195	194	158	115	7	0	37
Month 6	195	194	173	73	21	2	77

* Did not report; HS = high school.

Table A.7
Profile of Capital Guardian Youth ChalleNGe Academy (District of Columbia)

CAPITAL GUARDIAN YOUTH CHALLENGE ACADEMY, ESTABLISHED 2007					
Graduates Since Inception: 598			Program Type: GED		

Staffing

	Instructional	Cadre	Administrative	Case Managers	Recruiters	Total
Number employed	4	23	0	0	0	27

Funding

	Federal Funding	State Funding	Other Funding
Classes 48 and 49	$2,262,122	$2,143,109	$0

Residential Performance

	Dates	Applied	Entered Pre-ChalleNGe	Graduated	Received GED/HiSET	Received HS Credits	Received HS Diploma
Class 48	Jan. 2017– June 2017	125	34	29	0	0	6
Class 49	July 2017– Dec. 2017	101	71	45	0	0	12

Table A.7—Continued

Physical Fitness

	Push-Ups		1-Mile Run		BMI	
	Initial	Final	Initial	Final	Initial	Final
Class 48	29.9	33.4	13:11	11:44	25.4	*
Class 49	31.5	37.7	11:41	11:08	24.3	*

Responsible Citizenship

	Voting		Selective Service	
	Eligible	Registered	Eligible	Registered
Class 48	6	2	26	5
Class 49	26	26	33	11

Service to Community

	Service Hours per Cadet	Dollar Value per Hour	Total Value
Class 48	41	$39.45	$47,341
Class 49	56	$27.45	$99,006

Post-Residential Performance Status

	Graduated	Contacted	Placed	Education	Employment	Military	Multiple/ Other
Class 47							
Month 12	39	39	21	10	7	1	3
Class 48							
Month 1	29	28	23	2	19	0	2
Month 6	29	27	21	8	11	1	1
Month 12	29	25	20	9	7	1	3
Class 49							
Month 1	45	45	14	6	7	0	1
Month 6	45	40	28	10	15	1	2

* Did not report; HS = high school.

Table A.8
Profile of Florida Youth ChalleNGe Academy

FLORIDA YOUTH CHALLENGE ACADEMY, ESTABLISHED 2001						
Graduates Since Inception: 4,654			**Program Type: Credit Recovery, High School Diploma, GED**			

Staffing

	Instructional	Cadre	Administrative	Case Managers	Recruiters	Total
Number employed	11	41	9	5	2	68

Funding

	Federal Funding	State Funding	Other Funding
Classes 48 and 49	$4,019,500	$1,350,523	$291,057

Residential Performance

	Dates	Applied	Entered Pre-ChalleNGe	Graduated	Received GED/HiSET	Received HS Credits	Received HS Diploma
Class 48	Jan. 2017–June 2017	203	195	163	111	22	4
Class 49	July 2017–Dec. 2017	211	200	158	96	20	4

Physical Fitness

	Push-Ups		1-Mile Run		BMI	
	Initial	Final	Initial	Final	Initial	Final
Class 48	18.6	31.9	10:34	08:23	26.2	26.8
Class 49	16.5	31.0	10:53	08:25	*	25.1

Responsible Citizenship

	Voting		Selective Service	
	Eligible	Registered	Eligible	Registered
Class 48	36	36	31	31
Class 49	30	30	23	23

Service to Community

	Service Hours per Cadet	Dollar Value per Hour	Total Value
Class 48	60	$23.33	$226,646
Class 49	77	$23.33	$283,833

Table A.8—Continued

Post-Residential Performance Status

	Graduated	Contacted	Placed	Education	Employment	Military	Multiple/ Other
Class 47							
Month 12	149	82	65	13	39	6	7
Class 48							
Month 1	163	80	67	15	46	1	5
Month 6	163	77	61	17	34	1	9
Month 12	163	83	72	19	34	6	13
Class 49							
Month 1	158	84	58	21	31	1	5
Month 6	158	78	66	19	41	2	4

* Did not report; HS = high school.

Table A.9
Profile of Fort Gordon Youth ChalleNGe Academy (Georgia)

FORT GORDON YOUTH CHALLENGE ACADEMY, ESTABLISHED 2000

Graduates Since Inception: 6,069	Program Type: Credit Recovery, High School Diploma, GED

Staffing

	Instructional	Cadre	Administrative	Case Managers	Recruiters	Total
Number employed	8	38	4	4	1	55

Funding

	Federal Funding	State Funding	Other Funding
Classes 48 and 49	$10,597,770	$3,532,590	$0

Residential Performance

	Dates	Applied	Entered Pre-ChalleNGe	Graduated	Received GED/HiSET	Received HS Credits	Received HS Diploma
Class 48	March 2017– Aug. 2017	277	272	202	*	*	*
Class 49	Sept. 2017– Feb. 2018	255	214	135	*	*	*

Table A.9—Continued

Physical Fitness

	Push-Ups		1-Mile Run		BMI	
	Initial	Final	Initial	Final	Initial	Final
Class 48	30.0	42.0	09:36	09:21	*	*
Class 49	31.5	44.6	09:47	09:18	*	*

Responsible Citizenship

	Voting		Selective Service	
	Eligible	Registered	Eligible	Registered
Class 48	50	50	58	58
Class 49	34	34	69	69

Service to Community

	Service Hours per Cadet	Dollar Value per Hour	Total Value
Class 48	50	$25.15	$254,726
Class 49	47	$25.15	$160,697

Post-Residential Performance Status

	Graduated	Contacted	Placed	Education	Employment	Military	Multiple/Other
Class 47							
Month 12	146	127	95	26	68	0	1
Class 48							
Month 1	202	146	92	23	56	3	10
Month 6	202	146	79	16	60	0	4
Month 12	202	146	77	12	60	2	3
Class 49							
Month 1	135	75	38	14	21	0	3
Month 6	135	74	39	10	26	1	3

* Did not report; HS = high school.

Table A.10
Profile of Fort Stewart Youth ChalleNGe Academy (Georgia)

FORT STEWART YOUTH CHALLENGE ACADEMY, ESTABLISHED 1993					
Graduates Since Inception: 9,764			**Program Type: Credit Recovery, High School Diploma, GED**		

Staffing

	Instructional	Cadre	Administrative	Case Managers	Recruiters	Total
Number employed	6	42	38	5	2	93

Funding

	Federal Funding	State Funding	Other Funding
Classes 48 and 49	$5,666,300	$1,888,767	$0

Residential Performance

	Dates	Applied	Entered Pre-ChalleNGe	Graduated	Received GED/HiSET	Received HS Credits	Received HS Diploma
Class 48	Jan. 2017–June 2017	298	279	222	143	0	23
Class 49	July 2017–Dec. 2017	290	261	224	143	0	11

Physical Fitness

	Push-Ups		1-Mile Run		BMI	
	Initial	Final	Initial	Final	Initial	Final
Class 48	*	*	09:48	08:41	24.9	*
Class 49	29.0	46.1	09:51	08:19	23.7	23.8

Responsible Citizenship

	Voting		Selective Service	
	Eligible	Registered	Eligible	Registered
Class 48	55	55	44	44
Class 49	57	57	46	46

Service to Community

	Service Hours per Cadet	Dollar Value per Hour	Total Value
Class 48	65	$25.15	$364,701
Class 49	57	$25.15	$320,383

Table A.10—Continued

Post-Residential Performance Status

	Graduated	Contacted	Placed	Education	Employment	Military	Multiple/ Other
Class 47							
Month 12	225	187	171	56	78	10	27
Class 48							
Month 1	222	211	167	96	57	2	12
Month 6	222	207	183	110	62	4	7
Month 12	222	197	164	83	67	6	8
Class 49							
Month 1	224	213	151	98	36	5	12
Month 6	224	213	168	88	66	7	7

* Did not report; HS = high school.

Table A.11
Profile of Milledgeville Youth ChalleNGe Academy (Georgia)

MILLEDGEVILLE YOUTH CHALLENGE ACADEMY, ESTABLISHED 2016

Graduates Since Inception: 183	Program Type: Credit Recovery, High School Diploma, GED

Staffing

	Instructional	Cadre	Administrative	Case Managers	Recruiters	Total
Number employed	7	30	27	5	1	70

Funding

	Federal Funding	State Funding	Other Funding
Classes 48 and 49	$4,215,106	$1,405,035	$0

Residential Performance

	Dates	Applied	Entered Pre-ChalleNGe	Graduated	Received GED/HiSET	Received HS Credits	Received HS Diploma
Class 48	Oct. 2016–March 2017	168	168	76	32	10	1
Class 49	May 2017–Oct. 2017	205	160	107	44	11	3

Table A.11—Continued

Physical Fitness

	Push-Ups		1-Mile Run		BMI	
	Initial	Final	Initial	Final	Initial	Final
Class 48	49.3	54.2	06:16	06:29	24.9	*
Class 49	43.1	50.7	08:38	07:54	23.6	23.1

Responsible Citizenship

	Voting		Selective Service	
	Eligible	Registered	Eligible	Registered
Class 48	20	20	15	15
Class 49	24	24	22	22

Service to Community

	Service Hours per Cadet	Dollar Value per Hour	Total Value
Class 48	45	$25.15	$86,013
Class 49	51	$25.15	$137,163

Post-Residential Performance Status

	Graduated	Contacted	Placed	Education	Employment	Military	Multiple/ Other
Class 47							
Month 12	N/A	N/A	N/A	N/A	N/A	N/A	N/A
Class 48							
Month 1	76	70	40	17	17	0	7
Month 6	76	65	47	12	17	1	26
Month 12	76	63	47	9	27	2	20
Class 49							
Month 1	107	94	38	8	20	1	27
Month 6	107	94	69	13	34	3	35

* Did not report; HS = high school; N/A = not available.

Table A.12
Profile of Hawaii Youth ChalleNGe Academy at Barber's Point

HAWAII YOUTH CHALLENGE ACADEMY AT BARBER'S POINT, ESTABLISHED 1993					
Graduates Since Inception: 4,276			**Program Type: High School Diploma, HiSET**		

Staffing

	Instructional	Cadre	Administrative	Case Managers	Recruiters	Total
Number employed	6	18	6	1	2	33

Funding

	Federal Funding	State Funding	Other Funding
Classes 48 and 49	$3,187,500	$1,062,500	$0

Residential Performance

	Dates	Applied	Entered Pre-ChalleNGe	Graduated	Received GED/HiSET	Received HS Credits	Received HS Diploma
Class 48	Jan. 2017–June 2017	150	114	97	97	0	0
Class 49	July 2017–Dec. 2017	185	124	105	104	0	0

Physical Fitness

	Push-Ups		1-Mile Run		BMI	
	Initial	Final	Initial	Final	Initial	Final
Class 48	31.4	55.5	11:07	08:59	27.0	*
Class 49	23.5	50.6	11:08	08:27	27.0	*

Responsible Citizenship

	Voting		Selective Service	
	Eligible	Registered	Eligible	Registered
Class 48	18	16	39	39
Class 49	20	20	59	59

Service to Community

	Service Hours per Cadet	Dollar Value per Hour	Total Value
Class 48	116	$25.40	$285,801
Class 49	98	$25.40	$262,270

Table A.12—Continued

Post-Residential Performance Status

	Graduated	Contacted	Placed	Education	Employment	Military	Multiple/ Other
Class 47							
Month 12	121	107	73	3	65	2	3
Class 48							
Month 1	97	90	45	3	39	2	1
Month 6	97	87	54	6	39	4	5
Month 12	97	87	56	2	48	3	3
Class 49							
Month 1	105	87	57	22	18	1	16
Month 6	105	14	8	1	5	0	2

* Did not report; HS = high school.

Table A.13
Profile of Hawaii Youth ChalleNGe Academy at Hilo

HAWAII YOUTH CHALLENGE ACADEMY AT HILO, ESTABLISHED 2011		
Graduates Since Inception: 756		**Program Type: High School Diploma, HiSET**

Staffing

	Instructional	Cadre	Administrative	Case Managers	Recruiters	Total
Number employed	3	15	9	1	3	31

Funding

	Federal Funding	State Funding	Other Funding
Classes 48 and 49	$1,912,500	$637,500	$0

Residential Performance

	Dates	Applied	Entered Pre-ChalleNGe	Graduated	Received GED/HiSET	Received HS Credits	Received HS Diploma
Class 48	Jan. 2017–June 2017	79	71	62	0	0	62
Class 49	July 2017–Dec. 2017	105	88	67	0	0	67

Table A.13—Continued

Physical Fitness

	Push-Ups		1-Mile Run		BMI	
	Initial	Final	Initial	Final	Initial	Final
Class 48	33.3	60.8	08:33	07:42	25.8	25.3
Class 49	37.6	50.1	08:34	07:16	24.1	23.4

Responsible Citizenship

	Voting		Selective Service	
	Eligible	Registered	Eligible	Registered
Class 48	12	12	29	29
Class 49	12	12	31	31

Service to Community

	Service Hours per Cadet	Dollar Value per Hour	Total Value
Class 48	126	$25.40	$198,425
Class 49	120	$25.40	$204,216

Post-Residential Performance Status

	Graduated	Contacted	Placed	Education	Employment	Military	Multiple/ Other
Class 47							
Month 12	51	51	19	2	15	1	1
Class 48							
Month 1	62	62	22	4	17	0	2
Month 6	62	62	33	3	26	3	2
Month 12	62	62	30	0	25	4	2
Class 49							
Month 1	67	66	17	2	10	5	0
Month 6	67	66	27	5	15	7	0

* Did not report; HS = high school.

Table A.14
Profile of Idaho Youth ChalleNGe Academy

IDAHO YOUTH CHALLENGE ACADEMY, ESTABLISHED 2014		
Graduates Since Inception: 763	Program Type: Credit Recovery, High School Diploma, GED	

Staffing

	Instructional	Cadre	Administrative	Case Managers	Recruiters	Total
Number employed	7	21	20	4	2	54

Funding

	Federal Funding	State Funding	Other Funding
Classes 48 and 49	$2,646,901	$879,542	$593,034

Residential Performance

	Dates	Applied	Entered Pre-ChalleNGe	Graduated	Received GED/HiSET	Received HS Credits	Received HS Diploma
Class 48	Jan. 2017–June 2017	165	122	104	0	93	11
Class 49	July 2017–Dec. 2017	189	128	115	0	99	16

Physical Fitness

	Push-Ups		1-Mile Run		BMI	
	Initial	Final	Initial	Final	Initial	Final
Class 48	20.1	41.3	09:58	07:53	25.6	24.5
Class 49	19.0	43.6	09:38	08:03	24.4	23.6

Responsible Citizenship

	Voting		Selective Service	
	Eligible	Registered	Eligible	Registered
Class 48	20	20	30	30
Class 49	22	22	34	34

Service to Community

	Service Hours per Cadet	Dollar Value per Hour	Total Value
Class 48	46	$21.83	$103,527
Class 49	46	$21.83	$115,104

Table A.14—Continued

Post-Residential Performance Status

	Graduated	Contacted	Placed	Education	Employment	Military	Multiple/ Other
Class 47							
Month 12	104	79	66	16	12	4	34
Class 48							
Month 1	104	86	36	10	15	0	37
Month 6	104	88	68	25	8	1	40
Month 12	104	76	47	8	9	1	43
Class 49							
Month 1	115	106	44	37	4	0	25
Month 6	115	105	77	32	17	4	42

* Did not report; HS = high school.

Table A.15
Profile of Lincoln's ChalleNGe Academy (Illinois)

LINCOLN'S CHALLENGE ACADEMY, ESTABLISHED 1993	
Graduates Since Inception: 15,162	Program Type: GED

Staffing

	Instructional	Cadre	Administrative	Case Managers	Recruiters	Total
Number employed	7	41	38	5	6	97

Funding

	Federal Funding	State Funding	Other Funding
Classes 48 and 49	$5,284,000	$1,761,458	$1,003,667

Residential Performance

	Dates	Applied	Entered Pre-ChalleNGe	Graduated	Received GED/HiSET	Received HS Credits	Received HS Diploma
Class 48	Jan. 2017– June 2017	278	212	124	81	0	0
Class 49	July 2017– Dec. 2017	364	279	138	75	0	0

Table A.15—Continued

Physical Fitness

	Push-Ups		1-Mile Run		BMI	
	Initial	Final	Initial	Final	Initial	Final
Class 48	20.3	45.3	11:20	07:50	*	*
Class 49	20.2	48.4	10:26	09:11	*	*

Responsible Citizenship

	Voting		Selective Service	
	Eligible	Registered	Eligible	Registered
Class 48	27	0	23	23
Class 49	29	0	23	23

Service to Community

	Service Hours per Cadet	Dollar Value per Hour	Total Value
Class 48	61	$26.02	$195,880
Class 49	68	$26.02	$245,859

Post-Residential Performance Status

	Graduated	Contacted	Placed	Education	Employment	Military	Multiple/Other
Class 47							
Month 12	159	158	46	2	31	8	5
Class 48							
Month 1	124	31	31	5	23	1	2
Month 6	124	41	41	11	19	3	8
Month 12	124	19	19	1	12	4	2
Class 49							
Month 1	138	27	27	5	14	1	7
Month 6	138	23	23	0	17	2	4

* Did not report; HS = high school.

Table A.16
Profile of Hoosier Youth ChalleNGe Academy (Indiana)

HOOSIER YOUTH CHALLENGE ACADEMY, ESTABLISHED 2007		
Graduates Since Inception: 1,638		**Program Type: GED**

Staffing

	Instructional	Cadre	Administrative	Case Managers	Recruiters	Total
Number employed	4	11	13	7	3	38

Funding

	Federal Funding	State Funding	Other Funding
Classes 48 and 49	$3,009,506	$1,003,168	$28,772

Residential Performance

	Dates	Applied	Entered Pre-ChalleNGe	Graduated	Received GED/HiSET	Received HS Credits	Received HS Diploma
Class 48	Jan. 2017–June 2017	*	136	104	63	0	0
Class 49	July 2017–Dec. 2017	*	124	97	53	0	0

Physical Fitness

	Push-Ups		1-Mile Run		BMI	
	Initial	Final	Initial	Final	Initial	Final
Class 48	*	*	10:45	08:09	26.8	25.7
Class 49	24.4	35.6	10:47	08:47	24.8	25.3

Responsible Citizenship

	Voting		Selective Service	
	Eligible	Registered	Eligible	Registered
Class 48	15	15	40	40
Class 49	14	14	33	33

Service to Community

	Service Hours per Cadet	Dollar Value per Hour	Total Value
Class 48	40	$23.73	$98,717
Class 49	40	$23.73	$92,072

Table A.16—Continued

Post-Residential Performance Status

	Graduated	Contacted	Placed	Education	Employment	Military	Multiple/ Other
Class 47							
Month 12	90	88	3	0	0	0	3
Class 48							
Month 1	104	104	4	0	4	0	0
Month 6	104	104	1	0	1	0	0
Month 12	104	104	5	0	5	0	0
Class 49							
Month 1	97	97	23	5	16	0	2
Month 6	97	97	23	7	12	0	4

* Did not report; HS = high school.

Table A.17
Profile of Bluegrass ChalleNGe Academy (Kentucky)

BLUEGRASS CHALLENGE ACADEMY, ESTABLISHED 1993		
Graduates Since Inception: 3,113		Program Type: GED

Staffing

	Instructional	Cadre	Administrative	Case Managers	Recruiters	Total
Number employed	4	25	9	2	4	44

Funding

	Federal Funding	State Funding	Other Funding
Classes 48 and 49	$3,400,000	$259,000	$0

Residential Performance

	Dates	Applied	Entered Pre-ChalleNGe	Graduated	Received GED/HiSET	Received HS Credits	Received HS Diploma
Class 48	April 2017–Sept. 2017	147	147	96	*	*	*
Class 49	Oct. 2017–March 2018	136	136	72	*	*	*

Table A.17—Continued

Physical Fitness

	Push-Ups		1-Mile Run		BMI	
	Initial	Final	Initial	Final	Initial	Final
Class 48	27.1	50.2	10:59	09:46	25.7	*
Class 49	19.3	35.7	13:08	10:02	25.3	*

Responsible Citizenship

	Voting		Selective Service	
	Eligible	Registered	Eligible	Registered
Class 48	13	13	13	13
Class 49	14	14	14	14

Service to Community

	Service Hours per Cadet	Dollar Value per Hour	Total Value
Class 48	58	$21.17	$118,161
Class 49	55	$21.17	$84,068

Post-Residential Performance Status

	Graduated	Contacted	Placed	Education	Employment	Military	Multiple/ Other
Class 47							
Month 12	73	60	60	43	15	2	0
Class 48							
Month 1	96	96	93	84	6	0	3
Month 6	96	96	90	81	8	1	2
Month 12	96	N/A	N/A	N/A	N/A	N/A	N/A
Class 49							
Month 1	72	73	70	63	6	0	1
Month 6	72	N/A	N/A	N/A	N/A	N/A	N/A

* Did not report; HS = high school; N/A = not available.

Table A.18
Profile of Appalachian ChalleNGe Program (Kentucky)

APPALACHIAN CHALLENGE PROGRAM, ESTABLISHED 2012					
Graduates Since Inception: 884			**Program Type: Credit Recovery, GED**		

Staffing

	Instructional	Cadre	Administrative	Case Managers	Recruiters	Total
Number employed	5	26	11	3	3	48

Funding

	Federal Funding	State Funding	Other Funding
Classes 48 and 49	$2,634,081	$878,027	$0

Residential Performance

	Dates	Applied	Entered Pre-ChalleNGe	Graduated	Received GED/HiSET	Received HS Credits	Received HS Diploma
Class 48	Jan. 2017–June 2017	149	119	77	0	75	0
Class 49	July 2017–Dec. 2017	143	113	82	0	78	0

Physical Fitness

	Push-Ups		1-Mile Run		BMI	
	Initial	Final	Initial	Final	Initial	Final
Class 48	34.4	*	09:27	*	25.0	25.0
Class 49	28.6	44.6	09:57	09:13	24.5	24.5

Responsible Citizenship

	Voting		Selective Service	
	Eligible	Registered	Eligible	Registered
Class 48	13	13	30	30
Class 49	9	9	30	30

Service to Community

	Service Hours per Cadet	Dollar Value per Hour	Total Value
Class 48	67	$21.17	$109,216
Class 49	72	$21.17	$124,988

Table A.18—Continued

Post-Residential Performance Status

	Graduated	Contacted	Placed	Education	Employment	Military	Multiple/ Other*
Class 47							
Month 12	75	75	70	47	14	0	9
Class 48							
Month 1	77	68	64	52	6	1	5
Month 6	77	74	71	59	6	1	5
Month 12	77	74	72	60	6	1	5
Class 49							
Month 1	82	69	57	48	7	2	0
Month 6	82	70	68	59	7	2	0

* Did not report; HS = high school.

Table A.19
Profile of Louisiana Youth ChalleNGe Program—Camp Beauregard

LOUISIANA YOUTH CHALLENGE PROGRAM—CAMP BEAUREGARD, ESTABLISHED 1993		
Graduates Since Inception: 10,112		Program Type: Credit Recovery, HiSET

Staffing

	Instructional	Cadre	Administrative	Case Managers	Recruiters	Total
Number employed	15	46	46	12	2	121

Funding

	Federal Funding	State Funding	Other Funding
Classes 48 and 49	$6,375,000	$2,125,000	$0

Residential Performance

	Dates	Applied	Entered Pre-ChalleNGe	Graduated	Received GED/HiSET	Received HS Credits	Received HS Diploma
Class 48	Jan. 2017–June 2017	428	271	202	63	0	0
Class 49	July 2017–Dec. 2017	478	329	250	84	0	0

Table A.19—Continued

Physical Fitness

	Push-Ups		1-Mile Run		BMI	
	Initial	Final	Initial	Final	Initial	Final
Class 48	25.4	36.5	09:29	10:52	24.6	*
Class 49	22.4	34.7	09:24	08:53	24.3	*

Responsible Citizenship

	Voting		Selective Service	
	Eligible	Registered	Eligible	Registered
Class 48	34	34	90	90
Class 49	33	33	111	111

Service to Community

	Service Hours per Cadet	Dollar Value per Hour	Total Value
Class 48	44	$22.30	$199,689
Class 49	46	$22.30	$254,945

Post-Residential Performance Status

	Graduated	Contacted	Placed	Education	Employment	Military	Multiple/Other
Class 47							
Month 12	251	209	181	27	113	4	37
Class 48							
Month 1	202	200	186	26	99	0	64
Month 6	202	197	161	39	81	0	42
Month 12	202	189	162	18	95	1	48
Class 49							
Month 1	250	242	223	47	114	0	62
Month 6	250	222	193	28	98	5	63

* Did not report; HS = high school.

Table A.20
Profile of Louisiana Youth ChalleNGe Program—Camp Minden

LOUISIANA YOUTH CHALLENGE PROGRAM—CAMP MINDEN, ESTABLISHED 2002	
Graduates Since Inception: 5,081	Program Type: Credit Recovery, HiSET

Staffing

	Instructional	Cadre	Administrative	Case Managers	Recruiters	Total
Number employed	15	44	40	11	0	110

Funding

	Federal Funding	State Funding	Other Funding
Classes 48 and 49	$5,100,000	$1,700,000	$0

Residential Performance

	Dates	Applied	Entered Pre-ChalleNGe	Graduated	Received GED/HiSET	Received HS Credits	Received HS Diploma
Class 48	Feb. 2017–July 2017	333	251	207	77	0	0
Class 49	Aug. 2017–Jan. 2018	341	255	213	65	0	0

Physical Fitness

	Push-Ups		1-Mile Run		BMI	
	Initial	Final	Initial	Final	Initial	Final
Class 48	21.1	35.2	10:43	08:30	24.0	*
Class 49	24.0	35.0	09:42	08:03	25.0	*

Responsible Citizenship

	Voting		Selective Service	
	Eligible	Registered	Eligible	Registered
Class 48	37	0	73	73
Class 49	41	0	66	65

Service to Community

	Service Hours per Cadet	Dollar Value per Hour	Total Value
Class 48	56	$22.30	$259,051
Class 49	50	$22.30	$236,650

Table A.20—Continued

Post-Residential Performance Status

	Graduated	Contacted	Placed	Education	Employment	Military	Multiple/ Other
Class 47							
Month 12	235	232	203	44	110	2	47
Class 48							
Month 1	207	206	185	56	42	3	84
Month 6	207	205	185	58	65	4	58
Month 12	207	205	180	38	71	11	60
Class 49							
Month 1	213	213	196	55	45	4	92
Month 6	213	213	195	55	67	8	65

* Did not report; HS = high school.

Table A.21
Profile of Louisiana Youth ChalleNGe Program—Gillis Long

LOUISIANA YOUTH CHALLENGE PROGRAM—GILLIS LONG, ESTABLISHED 1999					
Graduates Since Inception: 8,184			Program Type: Credit Recovery, HiSET		

Staffing

	Instructional	Cadre	Administrative	Case Managers	Recruiters	Total
Number employed	17	42	13	10	1	83

Funding

	Federal Funding	State Funding	Other Funding
Classes 48 and 49	$6,515,625	$2,171,875	$0

Residential Performance

	Dates	Applied	Entered Pre-ChalleNGe	Graduated	Received GED/HiSET	Received HS Credits	Received HS Diploma
Class 48	April 2017–Sept. 2017	464	362	254	92	0	0
Class 49	Oct. 2017–March 2018	436	342	214	87	0	0

Table A.21—Continued

Physical Fitness

	Push-Ups		1-Mile Run		BMI	
	Initial	Final	Initial	Final	Initial	Final
Class 48	24.4	48.4	10:00	08:20	*	*
Class 49	27.0	38.3	11:01	10:25	*	*

Responsible Citizenship

	Voting		Selective Service	
	Eligible	Registered	Eligible	Registered
Class 48	81	35	34	34
Class 49	43	43	27	27

Service to Community

	Service Hours per Cadet	Dollar Value per Hour	Total Value
Class 48	54	$22.30	$305,867
Class 49	51	$22.30	$243,382

Post-Residential Performance Status

	Graduated	Contacted	Placed	Education	Employment	Military	Multiple/ Other
Class 47							
Month 12	258	226	211	51	85	4	71
Class 48							
Month 1	254	233	222	56	84	4	78
Month 6	254	223	215	64	89	3	59
Month 12	254	N/A	N/A	N/A	N/A	N/A	N/A
Class 49							
Month 1	214	188	172	36	84	5	47
Month 6	214	N/A	N/A	N/A	N/A	N/A	N/A

* Did not report; HS = high school; N/A = not available.

Table A.22
Profile of Freestate ChalleNGe Academy (Maryland)

FREESTATE CHALLENGE ACADEMY, ESTABLISHED 1993					
Graduates Since Inception: 4,284			Program Type: High School Diploma		

Staffing

	Instructional	Cadre	Administrative	Case Managers	Recruiters	Total
Number employed	4	43	18	7	2	74

Funding

	Federal Funding	State Funding	Other Funding
Classes 48 and 49	$3,320,900	$1,287,863	$150,000

Residential Performance

	Dates	Applied	Entered Pre-ChalleNGe	Graduated	Received GED/HiSET	Received HS Credits	Received HS Diploma
Class 48	Jan. 2017–June 2017	220	154	100	0	0	45
Class 49	July 2017–Dec. 2017	253	163	102	0	0	56

Physical Fitness

	Push-Ups		1-Mile Run		BMI	
	Initial	Final	Initial	Final	Initial	Final
Class 48	22.9	46.5	11:09	08:59	25.0	24.5
Class 49	23.9	33.1	10:51	09:36	26.9	25.0

Responsible Citizenship

	Voting		Selective Service	
	Eligible	Registered	Eligible	Registered
Class 48	26	26	41	41
Class 49	22	22	34	34

Service to Community

	Service Hours per Cadet	Dollar Value per Hour	Total Value
Class 48	52	$27.50	$143,633
Class 49	51	$27.50	$142,606

Table A.22—Continued

Post-Residential Performance Status

	Graduated	Contacted	Placed	Education	Employment	Military	Multiple/ Other
Class 47							
Month 12	103	103	60	11	44	1	4
Class 48							
Month 1	100	100	38	6	32	0	0
Month 6	100	88	45	15	28	1	1
Month 12	100	92	68	12	50	1	5
Class 49							
Month 1	102	96	19	4	15	0	0
Month 6	102	98	73	14	51	1	7

* Did not report; HS = high school.

Table A.23
Profile of Michigan Youth ChalleNGe Academy

MICHIGAN YOUTH CHALLENGE ACADEMY, ESTABLISHED 1999		
Graduates Since Inception: 3,587	Program Type: Credit Recovery, High School Diploma, GED	

Staffing

	Instructional	Cadre	Administrative	Case Managers	Recruiters	Total
Number employed	10	28	9	4	2	53

Funding

	Federal Funding	State Funding	Other Funding
Classes 48 and 49	$3,363,488	$1,121,163	$0

Residential Performance

	Dates	Applied	Entered Pre-ChalleNGe	Graduated	Received GED/HiSET	Received HS Credits	Received HS Diploma
Class 48	Jan. 2017– June 2017	195	144	116	0	107	9
Class 49	July 2017– Dec. 2017	248	155	119	0	94	25

Table A.23—Continued

Physical Fitness

	Push-Ups		1-Mile Run		BMI	
	Initial	Final	Initial	Final	Initial	Final
Class 48	33.8	51.6	09:31	08:04	*	*
Class 49	28.2	49.6	09:21	08:13	25.4	*

Responsible Citizenship

	Voting		Selective Service	
	Eligible	Registered	Eligible	Registered
Class 48	25	25	32	32
Class 49	20	20	23	23

Service to Community

	Service Hours per Cadet	Dollar Value per Hour	Total Value
Class 48	63	$23.91	$174,915
Class 49	63	$23.91	$178,943

Post-Residential Performance Status

	Graduated	Contacted	Placed	Education	Employment	Military	Multiple/Other
Class 47							
Month 12	112	111	86	18	37	7	24
Class 48							
Month 1	116	76	72	29	26	0	17
Month 6	116	88	79	26	25	6	22
Month 12	116	83	76	19	32	3	23
Class 49							
Month 1	119	99	92	60	15	0	17
Month 6	119	86	80	13	41	2	28

* Did not report; HS = high school.

Table A.24
Profile of Mississippi Youth ChalleNGe Academy

MISSISSIPPI YOUTH CHALLENGE ACADEMY, ESTABLISHED 1994		
Graduates Since Inception: 9,198		**Program Type: High School Diploma**

Staffing

	Instructional	Cadre	Administrative	Case Managers	Recruiters	Total
Number employed	10	47	26	7	4	94

Funding

	Federal Funding	State Funding	Other Funding
Classes 48 and 49	$4,350,000	$1,450,000	$0

Residential Performance

	Dates	Applied	Entered Pre-ChalleNGe	Graduated	Received GED/HiSET	Received HS Credits	Received HS Diploma
Class 48	Jan. 2017–June 2017	444	287	169	0	0	103
Class 49	July 2017–Dec. 2017	577	305	200	0	0	108

Physical Fitness

	Push-Ups		1-Mile Run		BMI	
	Initial	Final	Initial	Final	Initial	Final
Class 48	25.3	50.0	11:23	07:50	24.9	*
Class 49	22.6	48.3	11:59	07:59	25.1	*

Responsible Citizenship

	Voting		Selective Service	
	Eligible	Registered	Eligible	Registered
Class 48	45	45	60	60
Class 49	62	62	88	88

Service to Community

	Service Hours per Cadet	Dollar Value per Hour	Total Value
Class 48	73	$19.81	$244,962
Class 49	65	$19.81	$255,838

Table A.24—Continued

Post-Residential Performance Status

	Graduated	Contacted	Placed	Education	Employment	Military	Multiple/ Other
Class 47							
Month 12	211	170	164	27	91	11	35
Class 48							
Month 1	169	155	124	27	57	4	39
Month 6	169	142	129	23	66	8	35
Month 12	169	139	129	17	75	10	29
Class 49							
Month 1	200	192	148	31	61	7	54
Month 6	200	183	174	38	81	13	46

* Did not report; HS = high school.

Table A.25
Profile of Montana Youth ChalleNGe Academy

MONTANA YOUTH CHALLENGE ACADEMY, ESTABLISHED 1999					
Graduates Since Inception: 2,794			**Program Type: Credit Recovery, HiSET**		

Staffing

	Instructional	Cadre	Administrative	Case Managers	Recruiters	Total
Number employed	5	36	10	8	4	63

Funding

	Federal Funding	State Funding	Other Funding
Classes 48 and 49	$3,498,750	$1,166,250	$142,450

Residential Performance

	Dates	Applied	Entered Pre-ChalleNGe	Graduated	Received GED/HiSET	Received HS Credits	Received HS Diploma
Class 48	Jan. 2017–June 2017	137	111	80	31	0	0
Class 49	July 2017–Dec. 2017	127	90	80	29	0	0

Table A.25—Continued

Physical Fitness

	Push-Ups		1-Mile Run		BMI	
	Initial	Final	Initial	Final	Initial	Final
Class 48	*	*	12:09	08:05	24.6	*
Class 49	*	*	10:15	08:09	22.7	*

Responsible Citizenship

	Voting		Selective Service	
	Eligible	Registered	Eligible	Registered
Class 48	11	11	36	21
Class 49	14	14	42	19

Service to Community

	Service Hours per Cadet	Dollar Value per Hour	Total Value
Class 48	60	$22.42	$107,713
Class 49	58	$22.42	$104,526

Post-Residential Performance Status

	Graduated	Contacted	Placed	Education	Employment	Military	Multiple/Other
Class 47							
Month 12	95	79	78	10	20	1	47
Class 48							
Month 1	80	72	53	9	35	1	14
Month 6	80	70	61	25	27	1	11
Month 12	80	69	60	11	36	5	8
Class 49							
Month 1	80	71	52	31	16	0	11
Month 6	80	70	56	20	27	0	11

* Did not report; HS = high school.

Table A.26
Profile of Tarheel ChalleNGe Academy—New London (North Carolina)

TARHEEL CHALLENGE ACADEMY—NEW LONDON, ESTABLISHED 2015

Graduates Since Inception: 360	Program Type: Credit Recovery, High School Diploma, HiSET, GED

Staffing

	Instructional	Cadre	Administrative	Case Managers	Recruiters	Total
Number employed	7	21	21	3	2	54

Funding

	Federal Funding	State Funding	Other Funding
Classes 48 and 49	$2,622,979	$874,327	$191,898

Residential Performance

	Dates	Applied	Entered Pre-ChalleNGe	Graduated	Received GED/HiSET	Received HS Credits	Received HS Diploma
Class 48	May 2017–Oct. 2017	246	131	95	33	3	0
Class 49	Oct. 2017–April 2018	366	153	99	24	0	24

Physical Fitness

	Push-Ups		1-Mile Run		BMI	
	Initial	Final	Initial	Final	Initial	Final
Class 48	17.0	34.2	13:02	11:01	24.4	23.7
Class 49	22.5	37.1	13:00	10:59	25.1	23.8

Responsible Citizenship

	Voting		Selective Service	
	Eligible	Registered	Eligible	Registered
Class 48	11	11	14	14
Class 49	21	21	18	18

Service to Community

	Service Hours per Cadet	Dollar Value per Hour	Total Value
Class 48	50	$23.41	$110,308
Class 49	90	$23.41	$207,424

Table A.26—Continued

Post-Residential Performance Status

	Graduated	Contacted	Placed	Education	Employment	Military	Multiple/ Other
Class 47							
Month 12	50	45	25	8	10	0	7
Class 48							
Month 1	95	93	36	8	19	0	9
Month 6	95	95	27	19	7	0	1
Month 12	95	N/A	N/A	N/A	N/A	N/A	N/A
Class 49							
Month 1	99	99	35	10	17	1	7
Month 6	99	N/A	N/A	N/A	N/A	N/A	N/A

* Did not report; HS = high school; N/A = not available.

Table A.27
Profile of Tarheel ChalleNGe Academy—Salemburg (North Carolina)

TARHEEL CHALLENGE ACADEMY—SALEMBURG, ESTABLISHED 1994

Graduates Since Inception: 4,713   Program Type: Credit Recovery, High School Diploma, HiSET, GED

Staffing

	Instructional	Cadre	Administrative	Case Managers	Recruiters	Total
Number employed	10	32	22	3	2	69

Funding

	Federal Funding	State Funding	Other Funding
Classes 48 and 49	$2,625,000	$875,000	$0

Residential Performance

	Dates	Applied	Entered Pre-ChalleNGe	Graduated	Received GED/HiSET	Received HS Credits	Received HS Diploma
Class 48	Jan. 2017– June 2017	261	147	88	54	0	0
Class 49	July 2017– Dec. 2017	289	156	105	*	*	*

Table A.27—Continued

Physical Fitness

	Push-Ups		1-Mile Run		BMI	
	Initial	Final	Initial	Final	Initial	Final
Class 48	26.2	42.2	10:34	07:52	23.6	24.2
Class 49	15.4	34.4	11:15	08:28	25.4	24.5

Responsible Citizenship

	Voting		Selective Service	
	Eligible	Registered	Eligible	Registered
Class 48	46	46	10	10
Class 49	61	61	25	25

Service to Community

	Service Hours per Cadet	Dollar Value per Hour	Total Value
Class 48	100	$23.41	$206,214
Class 49	83	$23.41	$204,510

Post-Residential Performance Status

	Graduated	Contacted	Placed	Education	Employment	Military	Multiple/ Other
Class 47							
Month 12	112	5	3	0	3	0	0
Class 48							
Month 1	88	69	64	6	29	0	29
Month 6	88	88	75	14	20	2	47
Month 12	88	88	60	7	11	2	48
Class 49							
Month 1	105	105	30	3	6	1	20
Month 6	105	105	78	17	30	4	27

* Did not report; HS = high school.

Table A.28
Profile of New Jersey Youth ChalleNGe Academy

NEW JERSEY YOUTH CHALLENGE ACADEMY, ESTABLISHED 1994					
Graduates Since Inception: 3,931			**Program Type: High School Diploma, GED**		

Staffing

	Instructional	Cadre	Administrative	Case Managers	Recruiters	Total
Number employed	5	28	9	2	4	48

Funding

	Federal Funding	State Funding	Other Funding
Classes 48 and 49	$2,700,000	$900,000	$450,201

Residential Performance

	Dates	Applied	Entered Pre-ChalleNGe	Graduated	Received GED/HiSET	Received HS Credits	Received HS Diploma
Class 48	April 2017–Sept. 2017	286	141	82	0	0	29
Class 49	Oct. 2017–March 2018	254	122	60	0	0	24

Physical Fitness

	Push-Ups		1-Mile Run		BMI	
	Initial	Final	Initial	Final	Initial	Final
Class 48	29.6	55.9	11:03	08:34	25.6	25.0
Class 49	29.5	46.0	10:36	11:42	25.6	25.3

Responsible Citizenship

	Voting		Selective Service	
	Eligible	Registered	Eligible	Registered
Class 48	24	24	19	19
Class 49	13	13	8	8

Service to Community

	Service Hours per Cadet	Dollar Value per Hour	Total Value
Class 48	48	$28.32	$111,235
Class 49	48	$28.32	$81,137

Table A.28—Continued

Post-Residential Performance Status

	Graduated	Contacted	Placed	Education	Employment	Military	Multiple/ Other
Class 47							
Month 12	81	81	79	28	30	9	12
Class 48							
Month 1	82	82	28	10	12	0	6
Month 6	82	82	68	23	27	3	15
Month 12	82	82	76	27	28	4	17
Class 49							
Month 1	60	60	18	6	8	2	2
Month 6	60	60	44	14	21	2	7

* Did not report; HS = high school.

Table A.29
Profile of New Mexico Youth ChalleNGe Academy

NEW MEXICO YOUTH CHALLENGE ACADEMY, ESTABLISHED 2001					
Graduates Since Inception: 2,545			Program Type: HiSET		

Staffing

	Instructional	Cadre	Administrative	Case Managers	Recruiters	Total
Number employed	0	30	12	3	3	48

Funding

	Federal Funding	State Funding	Other Funding
Classes 48 and 49	$2,403,980	$800,000	$225,000

Residential Performance

	Dates	Applied	Entered Pre-ChalleNGe	Graduated	Received GED/HiSET	Received HS Credits	Received HS Diploma
Class 48	Jan. 2017–June 2017	156	119	93	61	0	0
Class 49	July 2017–Dec. 2017	168	148	107	63	0	0

Table A.29—Continued

Physical Fitness

	Push-Ups		1-Mile Run		BMI	
	Initial	Final	Initial	Final	Initial	Final
Class 48	33.2	55.1	09:56	06:37	25.0	*
Class 49	21.5	43.8	09:43	06:47	25.5	*

Responsible Citizenship

	Voting		Selective Service	
	Eligible	Registered	Eligible	Registered
Class 48	23	23	39	39
Class 49	22	22	46	46

Service to Community

	Service Hours per Cadet	Dollar Value per Hour	Total Value
Class 48	80	$20.58	$152,827
Class 49	45	$20.58	$98,588

Post-Residential Performance Status

	Graduated	Contacted	Placed	Education	Employment	Military	Multiple/Other
Class 47							
Month 12	109	86	61	8	39	3	11
Class 48							
Month 1	93	93	93	4	31	0	58
Month 6	93	82	82	3	34	1	44
Month 12	93	10	10	1	3	0	6
Class 49							
Month 1	107	85	85	9	21	2	53
Month 6	107	9	9	0	2	0	7

* Did not report; HS = high school.

Table A.30
Profile of Thunderbird Youth Academy (Oklahoma)

THUNDERBIRD YOUTH ACADEMY, ESTABLISHED 1993					

Graduates Since Inception: 4,637			Program Type: Credit Recovery, High School Diploma, GED		

Staffing

	Instructional	Cadre	Administrative	Case Managers	Recruiters	Total
Number employed	6	49	14	4	6	79

Funding

	Federal Funding	State Funding	Other Funding
Classes 48 and 49	$2,878,193	$959,398	$49,000

Residential Performance

	Dates	Applied	Entered Pre-ChalleNGe	Graduated	Received GED/HiSET	Received HS Credits	Received HS Diploma
Class 48	Jan. 2017–June 2017	374	151	101	0	68	31
Class 49	July 2017–Dec. 2017	451	174	114	6	83	24

Physical Fitness

	Push-Ups		1-Mile Run		BMI	
	Initial	Final	Initial	Final	Initial	Final
Class 48	38.5	42.1	10:55	09:13	25.2	24.7
Class 49	28.4	44.9	09:54	08:48	25.4	25.7

Responsible Citizenship

	Voting		Selective Service	
	Eligible	Registered	Eligible	Registered
Class 48	21	21	35	35
Class 49	16	16	27	27

Service to Community

	Service Hours per Cadet	Dollar Value per Hour	Total Value
Class 48	60	$22.18	$135,031
Class 49	49	$22.18	$122,899

Table A.30—Continued

Post-Residential Performance Status

	Graduated	Contacted	Placed	Education	Employment	Military	Multiple/Other
Class 47							
Month 12	99	88	79	27	19	5	28
Class 48							
Month 1	101	101	89	60	6	2	21
Month 6	101	101	92	48	8	5	31
Month 12	101	101	89	32	16	5	36
Class 49							
Month 1	114	114	102	80	4	9	9
Month 6	114	114	105	40	16	6	44

* Did not report; HS = high school.

Table A.31
Profile of Oregon Youth ChalleNGe Program

OREGON YOUTH CHALLENGE PROGRAM, ESTABLISHED 1999						
Graduates Since Inception: 4,376			Program Type: Credit Recovery, High School Diploma, GED			

Staffing

	Instructional	Cadre	Administrative	Case Managers	Recruiters	Total
Number employed	5	25	13	3	1	47

Funding

	Federal Funding	State Funding	Other Funding
Classes 48 and 49	$4,016,630	$1,336,800	$2,280

Residential Performance

	Dates	Applied	Entered Pre-ChalleNGe	Graduated	Received GED/HiSET	Received HS Credits	Received HS Diploma
Class 48	Jan. 2017–June 2017	293	156	131	0	118	13
Class 49	July 2017–Dec. 2017	198	156	123	0	107	16

Table A.31—Continued

Physical Fitness

	Push-Ups		1-Mile Run		BMI	
	Initial	Final	Initial	Final	Initial	Final
Class 48	17.6	29.7	09:56	07:22	26.3	25.4
Class 49	29.7	46.4	09:43	07:22	26.3	25.8

Responsible Citizenship

	Voting		Selective Service	
	Eligible	Registered	Eligible	Registered
Class 48	66	66	31	31
Class 49	76	76	39	39

Service to Community

	Service Hours per Cadet	Dollar Value per Hour	Total Value
Class 48	92	$24.89	$299,974
Class 49	93	$24.89	$284,717

Post-Residential Performance Status

	Graduated	Contacted	Placed	Education	Employment	Military	Multiple/Other
Class 47							
Month 12	135	120	108	46	41	10	11
Class 48							
Month 1	131	126	117	52	37	0	28
Month 6	131	107	101	79	19	1	4
Month 12	131	110	101	65	27	6	5
Class 49							
Month 1	123	113	106	77	14	1	16
Month 6	123	85	80	53	19	5	6

* Did not report; HS = high school.

Table A.32
Profile of Puerto Rico Youth ChalleNGe Academy

PUERTO RICO YOUTH CHALLENGE ACADEMY, ESTABLISHED 1999					
Graduates Since Inception: 5,749			Program Type: Credit Recovery, High School Diploma		

Staffing

	Instructional	Cadre	Administrative	Case Managers	Recruiters	Total
Number employed	12	47	35	10	1	105

Funding

	Federal Funding	State Funding	Other Funding
Classes 48 and 49	$3,500,000	$1,166,667	$650,000

Residential Performance

	Dates	Applied	Entered Pre-ChalleNGe	Graduated	Received GED/HiSET	Received HS Credits	Received HS Diploma
Class 48	April 2017– Sept. 2017	280	271	220	*	*	*
Class 49	Dec. 2017– May 2018	247	247	216	0	1	215

Physical Fitness

	Push-Ups		1-Mile Run		BMI	
	Initial	Final	Initial	Final	Initial	Final
Class 48	*	*	*	*	*	*
Class 49	24.3	37.8	09:43	08:07	24.0	23.2

Responsible Citizenship

	Voting		Selective Service	
	Eligible	Registered	Eligible	Registered
Class 48	37	36	35	35
Class 49	81	80	69	69

Service to Community

	Service Hours per Cadet	Dollar Value per Hour	Total Value
Class 48	70	$12.71	$195,734
Class 49	50	$12.71	$137,268

Table A.32—Continued

Post-Residential Performance Status							
	Graduated	Contacted	Placed	Education	Employment	Military	Multiple/ Other
Class 47							
Month 12	225	224	196	132	54	2	8
Class 48							
Month 1	220	57	57	29	19	0	9
Month 6	220	157	157	112	41	3	1
Month 12	220	N/A	N/A	N/A	N/A	N/A	N/A
Class 49							
Month 1	216	100	100	23	64	3	10
Month 6	216	N/A	N/A	N/A	N/A	N/A	N/A

* Did not report; HS = high school; N/A = not available.

Table A.33
Profile of South Carolina Youth ChalleNGe Academy

SOUTH CAROLINA YOUTH CHALLENGE ACADEMY, ESTABLISHED 1998					
Graduates Since Inception: 3,561			Program Type: GED		

Staffing						
	Instructional	Cadre	Administrative	Case Managers	Recruiters	Total
Number employed	5	35	24	4	1	69

Funding			
	Federal Funding	State Funding	Other Funding
Classes 48 and 49	$2,750,520	$914,667	$0

Residential Performance						
	Dates	Applied	Entered Pre-ChalleNGe	Graduated	Received GED/HiSET	Received HS Credits	Received HS Diploma
Class 48	Jan. 2017–June 2017	186	160	112	53	0	0
Class 49	July 2017–Dec. 2017	175	139	103	43	0	0

Table A.33—Continued

Physical Fitness

	Push-Ups		1-Mile Run		BMI	
	Initial	Final	Initial	Final	Initial	Final
Class 48	*	*	09:49	08:00	24.2	24.0
Class 49	*	*	10:02	07:58	24.7	23.1

Responsible Citizenship

	Voting		Selective Service	
	Eligible	Registered	Eligible	Registered
Class 48	25	9	18	7
Class 49	17	9	16	9

Service to Community

	Service Hours per Cadet	Dollar Value per Hour	Total Value
Class 48	40	$22.22	$99,546
Class 49	40	$22.22	$91,546

Post-Residential Performance Status

	Graduated	Contacted	Placed	Education	Employment	Military	Multiple/Other
Class 47							
Month 12	111	110	79	8	11	5	55
Class 48							
Month 1	112	80	45	31	14	0	0
Month 6	112	55	30	9	19	0	2
Month 12	112	57	20	3	13	1	3
Class 49							
Month 1	103	50	36	23	10	0	3
Month 6	103	50	25	7	14	1	3

* Did not report; HS = high school.

Table A.34
Profile of Volunteer ChalleNGe Academy (Tennessee)

VOLUNTEER CHALLENGE ACADEMY, ESTABLISHED 2017	
Graduates Since Inception: 23	Program Type: High School Diploma

Staffing

	Instructional	Cadre	Administrative	Case Managers	Recruiters	Total
Number employed	4	41	5	2	2	54

Funding

	Federal Funding	State Funding	Other Funding
Classes 48 and 49	$2,579,361	$859,787	$0

Residential Performance

	Dates	Applied	Entered Pre-ChalleNGe	Graduated	Received GED/HiSET	Received HS Credits	Received HS Diploma
Class 48	N/A	N/A	N/A	N/A	N/A	N/A	N/A
Class 49	July 2017–Dec. 2017	51	*	23	8	0	0

Physical Fitness

	Push-Ups		1-Mile Run		BMI	
	Initial	Final	Initial	Final	Initial	Final
Class 48	N/A	N/A	N/A	N/A	N/A	N/A
Class 49	37.9	46.0	10:40	09:49	*	*

Responsible Citizenship

	Voting		Selective Service	
	Eligible	Registered	Eligible	Registered
Class 48	N/A	N/A	N/A	N/A
Class 49	3	0	3	0

Service to Community

	Service Hours per Cadet	Dollar Value per Hour	Total Value
Class 48	N/A	N/A	N/A
Class 49	43	$21.98	$21,738

Table A.34—Continued

Post-Residential Performance Status

	Graduated	Contacted	Placed	Education	Employment	Military	Multiple/ Other
Class 47							
Month 12	N/A	N/A	N/A	N/A	N/A	N/A	N/A
Class 48							
Month 1	N/A	N/A	N/A	N/A	N/A	N/A	N/A
Month 6	N/A	N/A	N/A	N/A	N/A	N/A	N/A
Month 12	N/A	N/A	N/A	N/A	N/A	N/A	N/A
Class 49							
Month 1	23	16	15	6	6	2	1
Month 6	23	18	18	11	5	2	0

* Did not report; HS = high school; N/A = not available.

Table A.35
Profile of Texas ChalleNGe Academy—East

TEXAS CHALLENGE ACADEMY—EAST, ESTABLISHED 2014					
Graduates Since Inception: 298			**Program Type: Credit Recovery, High School Diploma, GED**		

Staffing

	Instructional	Cadre	Administrative	Case Managers	Recruiters	Total
Number employed	0	25	8	5	6	44

Funding

	Federal Funding	State Funding	Other Funding
Classes 48 and 49	$2,550,000	$850,000	$96,643

Residential Performance

	Dates	Applied	Entered Pre-ChalleNGe	Graduated	Received GED/HiSET	Received HS Credits	Received HS Diploma
Class 48	Jan. 2017–June 2017	219	141	79	0	64	15
Class 49	July 2017–Dec. 2017	224	118	55	2	40	12

Table A.35—Continued

Physical Fitness

	Push-Ups[a]		1-Mile Run		BMI	
	Initial	Final	Initial	Final	Initial	Final
Class 48	7.0	7.9	08:59	08:37	22.9	23.4
Class 49	6.6	6.1	12:12	09:45	*	*

Responsible Citizenship

	Voting		Selective Service	
	Eligible	Registered	Eligible	Registered
Class 48	19	17	11	11
Class 49	23	20	15	15

Service to Community

	Service Hours per Cadet	Dollar Value per Hour	Total Value
Class 48	48	$24.64	$93,240
Class 49	70	$24.64	$95,135

Post-Residential Performance Status

	Graduated	Contacted	Placed	Education	Employment	Military	Multiple/Other
Class 47							
Month 12	64	64	33	16	7	1	9
Class 48							
Month 1	79	79	24	6	16	1	1
Month 6	79	79	54	33	13	5	3
Month 12	79	79	37	17	13	4	3
Class 49							
Month 1	55	53	24	14	6	0	4
Month 6	55	53	25	12	10	2	1

* Did not report; HS = high school.

[a] Pull-ups; site does not collect data on push-ups.

Table A.36
Profile of Texas ChalleNGe Academy—West

TEXAS CHALLENGE ACADEMY—WEST, ESTABLISHED 1999

Graduates Since Inception: 3,157	Program Type: Credit Recovery, High School Diploma, GED

Staffing

	Instructional	Cadre	Administrative	Case Managers	Recruiters	Total
Number employed	4	14	5	4	4	31

Funding

	Federal Funding	State Funding	Other Funding
Classes 48 and 49	$2,550,000	$850,000	$253,100

Residential Performance

	Dates	Applied	Entered Pre-ChalleNGe	Graduated	Received GED/HiSET	Received HS Credits	Received HS Diploma
Class 48	March 2017–Aug. 2017	247	88	41	0	31	10
Class 49	Sept. 2017–Feb. 2018	145	106	61	0	52	9

Physical Fitness

	Push-Ups[a]		1-Mile Run		BMI	
	Initial	Final	Initial	Final	Initial	Final
Class 48	7.4	9.8	10:19	09:23	23.5	23.9
Class 49	8.5	54.0	10:29	08:09	25.0	24.7

Responsible Citizenship

	Voting		Selective Service	
	Eligible	Registered	Eligible	Registered
Class 48	8	8	8	8
Class 49	19	19	20	20

Service to Community

	Service Hours per Cadet	Dollar Value per Hour	Total Value
Class 48	40	$24.64	$40,410
Class 49	40	$24.64	$60,122

Table A.36—Continued

Post-Residential Performance Status

	Graduated	Contacted	Placed	Education	Employment	Military	Multiple/ Other
Class 47							
Month 12	78	78	42	20	20	2	0
Class 48							
Month 1	41	39	20	13	6	0	1
Month 6	41	34	30	15	13	2	0
Month 12	41	33	28	8	18	2	0
Class 49							
Month 1	61	26	26	13	13	0	0
Month 6	61	38	38	12	24	2	0

* Did not report; HS = high school.

[a] Site collected data on pull-ups instead of push-ups for Class 48. For Class 49, site initially measured pull-ups and then switched to push-ups for the final assessment.

Table A.37
Profile of Virginia Commonwealth ChalleNGe Youth Academy

VIRGINIA COMMONWEALTH CHALLENGE YOUTH ACADEMY, ESTABLISHED 1994

Graduates Since Inception: 4,801	Program Type: Credit Recovery, High School Diploma, GED

Staffing

	Instructional	Cadre	Administrative	Case Managers	Recruiters	Total
Number employed	9	32	15	4	3	63

Funding

	Federal Funding	State Funding	Other Funding
Classes 48 and 49	$4,004,000	$1,334,667	$0

Residential Performance

	Dates	Applied	Entered Pre-ChalleNGe	Graduated	Received GED/HiSET	Received HS Credits	Received HS Diploma
Class 48	Oct. 2016– Feb. 2017	240	170	100	39	0	0
Class 49	March 2017– Aug. 2017	221	186	116	41	0	0

Table A.37—Continued

Physical Fitness

	Push-Ups		1-Mile Run		BMI	
	Initial	**Final**	**Initial**	**Final**	**Initial**	**Final**
Class 48	25.8	35.3	10:52	09:52	24.9	*
Class 49	25.9	42.8	09:28	08:32	24.8	*

Responsible Citizenship

	Voting		Selective Service	
	Eligible	**Registered**	**Eligible**	**Registered**
Class 48	16	10	37	22
Class 49	23	22	47	46

Service to Community

	Service Hours per Cadet	**Dollar Value per Hour**	**Total Value**
Class 48	103	$26.75	$275,525
Class 49	157	$26.75	$487,171

Post-Residential Performance Status

	Graduated	**Contacted**	**Placed**	**Education**	**Employment**	**Military**	**Multiple/ Other**
Class 47							
Month 12	93	11	11	1	6	1	3
Class 48							
Month 1	100	45	39	8	20	0	11
Month 6	100	38	28	6	8	0	14
Month 12	100	35	34	14	8	1	11
Class 49							
Month 1	116	56	47	22	15	0	12
Month 6	116	45	36	16	13	2	5

* Did not report; HS = high school.

Table A.38
Profile of Washington Youth Academy

WASHINGTON YOUTH ACADEMY, ESTABLISHED 2009					
Graduates Since Inception: 2,186			**Program Type: Credit Recovery**		

Staffing

	Instructional	Cadre	Administrative	Case Managers	Recruiters	Total
Number employed	7	32	6	6	2	53

Funding

	Federal Funding	State Funding	Other Funding
Classes 48 and 49	$3,700,000	$1,233,333	$1,881,183

Residential Performance

	Dates	Applied	Entered Pre-ChalleNGe	Graduated	Received GED/HiSET	Received HS Credits	Received HS Diploma
Class 48	Jan. 2017–June 2017	243	161	143	0	139	0
Class 49	July 2017–Dec. 2017	296	167	141	0	137	0

Physical Fitness

	Push-Ups		1-Mile Run		BMI	
	Initial	Final	Initial	Final	Initial	Final
Class 48	25.4	40.1	10:40	07:39	26.9	26.1
Class 49	26.0	36.3	10:00	08:09	26.3	*

Responsible Citizenship

	Voting		Selective Service	
	Eligible	Registered	Eligible	Registered
Class 48	36	36	68	68
Class 49	34	34	70	70

Service to Community

	Service Hours per Cadet	Dollar Value per Hour	Total Value
Class 48	65	$30.46	$283,126
Class 49	67	$30.46	$287,756

Table A.38—Continued

Post-Residential Performance Status

	Graduated	Contacted	Placed	Education	Employment	Military	Multiple/ Other
Class 47							
Month 12	149	149	112	67	35	7	3
Class 48							
Month 1	143	116	63	48	10	0	7
Month 6	143	129	122	114	6	2	0
Month 12	143	129	116	102	12	2	1
Class 49							
Month 1	141	135	119	119	0	0	4
Month 6	141	130	118	115	0	1	4

* Did not report; HS = high school.

Table A.39
Profile of Wisconsin ChalleNGe Academy

WISCONSIN CHALLENGE ACADEMY, ESTABLISHED 1998					
Graduates Since Inception: 3,623			Program Type: Credit Recovery, High School Diploma, GED		

Staffing

	Instructional	Cadre	Administrative	Case Managers	Recruiters	Total
Number employed	5	25	6	5	4	45

Funding

	Federal Funding	State Funding	Other Funding
Classes 48 and 49	$3,523,900	$1,174,633	$1,200

Residential Performance

	Dates	Applied	Entered Pre-ChalleNGe	Graduated	Received GED/HiSET	Received HS Credits	Received HS Diploma
Class 48	Jan. 2017– June 2017	253	146	92	40	0	28
Class 49	July 2017– Dec. 2017	275	158	108	43	0	38

Table A.39—Continued

Physical Fitness

	Push-Ups		1-Mile Run		BMI	
	Initial	**Final**	**Initial**	**Final**	**Initial**	**Final**
Class 48	15.2	24.6	10:04	07:24	23.5	24.4
Class 49	10.6	25.7	09:14	07:50	24.7	25.8

Responsible Citizenship

	Voting		Selective Service	
	Eligible	**Registered**	**Eligible**	**Registered**
Class 48	28	27	50	50
Class 49	27	27	54	54

Service to Community

	Service Hours per Cadet	Dollar Value per Hour	Total Value
Class 48	59	$24.00	$130,272
Class 49	68	$24.00	$176,256

Post-Residential Performance Status

	Graduated	Contacted	Placed	Education	Employment	Military	Multiple/ Other
Class 47							
Month 12	116	114	92	9	43	5	35
Class 48							
Month 1	92	92	55	2	47	0	19
Month 6	92	90	63	3	41	3	26
Month 12	92	66	44	2	29	6	19
Class 49							
Month 1	108	108	66	16	26	0	43
Month 6	108	107	81	12	46	3	30

* Did not report; HS = high school.

Table A.40
Profile of Mountaineer ChalleNGe Academy (West Virginia)

MOUNTAINEER CHALLENGE ACADEMY, ESTABLISHED 1993

Graduates Since Inception: 4,046	Program Type: High School Diploma

Staffing

	Instructional	Cadre	Administrative	Case Managers	Recruiters	Total
Number employed	9	36	16	6	4	71

Funding

	Federal Funding	State Funding	Other Funding
Classes 48 and 49	$3,933,003	$1,311,001	$119,830

Residential Performance

	Dates	Applied	Entered Pre-ChalleNGe	Graduated	Received GED/HiSET	Received HS Credits	Received HS Diploma
Class 48	Jan. 2017–June 2017	364	189	151	0	20	129
Class 49	July 2017–Dec. 2017	348	194	153	0	27	126

Physical Fitness

	Push-Ups		1-Mile Run		BMI	
	Initial	Final	Initial	Final	Initial	Final
Class 48	*	*	10:12	07:30	24.5	23.6
Class 49	*	*	09:35	07:46	24.7	23.7

Responsible Citizenship

	Voting		Selective Service	
	Eligible	Registered	Eligible	Registered
Class 48	32	32	31	31
Class 49	31	31	29	29

Service to Community

	Service Hours per Cadet	Dollar Value per Hour	Total Value
Class 48	63	$21.10	$200,724
Class 49	62	$21.10	$200,155

Table A.40—Continued

Post-Residential Performance Status

	Graduated	Contacted	Placed	Education	Employment	Military	Multiple/ Other
Class 47							
Month 12	159	159	101	13	64	17	7
Class 48							
Month 1	151	140	25	0	24	0	1
Month 6	151	146	82	7	65	5	5
Month 12	151	115	93	6	74	11	2
Class 49							
Month 1	153	153	18	1	16	1	0
Month 6	153	127	95	3	79	9	4

* Did not report; HS = high school.

Table A.41
Profile of Wyoming Cowboy ChalleNGe Academy

WYOMING COWBOY CHALLENGE ACADEMY, ESTABLISHED 2005					
Graduates Since Inception: 1,029			Program Type: Credit Recovery, HiSET		

Staffing

	Instructional	Cadre	Administrative	Case Managers	Recruiters	Total
Number employed	4	22	12	2	5	45

Funding

	Federal Funding	State Funding	Other Funding
Classes 48 and 49	$2,000,000	$500,000	$2,390,000

Residential Performance

	Dates	Applied	Entered Pre-ChalleNGe	Graduated	Received GED/HiSET	Received HS Credits	Received HS Diploma
Class 48	Jan. 2017– June 2017	160	119	86	42	15	2
Class 49	July 2017– Dec. 2017	122	100	60	32	3	0

Table A.41—Continued

Physical Fitness

	Push-Ups		1-Mile Run		BMI	
	Initial	Final	Initial	Final	Initial	Final
Class 48	31.4	38.2	09:10	07:29	23.9	24.3
Class 49	18.7	25.1	10:46	09:02	24.5	25.1

Responsible Citizenship

	Voting		Selective Service	
	Eligible	Registered	Eligible	Registered
Class 48	16	16	13	13
Class 49	8	8	7	7

Service to Community

	Service Hours per Cadet	Dollar Value per Hour	Total Value
Class 48	41	$23.17	$82,317
Class 49	42	$23.17	$58,790

Post-Residential Performance Status

	Graduated	Contacted	Placed	Education	Employment	Military	Multiple/Other
Class 47							
Month 12	48	48	40	12	14	4	10
Class 48							
Month 1	86	49	32	11	21	0	0
Month 6	86	47	44	17	26	1	0
Month 12	86	62	58	21	31	3	3
Class 49							
Month 1	60	40	25	10	11	2	2
Month 6	60	30	28	14	9	3	2

* Did not report; HS = high school.

References

Bloom, Dan, Alissa Gardenhire-Crooks, and Conrad Mandsager, *Reengaging High School Dropouts: Early Results of the National Guard Youth ChalleNGe Program Evaluation*, New York: MDRC, February 2009.

Comprehensive Adult Student Assessment System, "Study of the CASAS Relationship to GED 2002," San Diego, Calif., research brief, June 2003. As of November 8, 2017:
https://www.casas.org/docs/default-source/research/download-what-is-the-relationship-between-casas-assessment-and-ged-2002-.pdf?sfvrsn=5?Status=Master

———, "Study of the CASAS Relationship to GED 2014," San Diego, Calif., research brief, March 2016. As of November 8, 2017:
https://www.casas.org/docs/default-source/research/study-of-the-casas-relationship-to-ged-2014.pdf

Corbin, Juliet, and Anselm Strauss, "Grounded Theory Research: Procedures, Canons and Evaluative Criteria," *Zeitschrift Für Soziologie*, Vol. 19, No. 6, 1990, 418–427.

Independent Sector, "Value of Volunteer Time," webpage, May 31, 2016. As of December 7, 2017:
http://www.independentsector.org/resource/the-value-of-volunteer-time/

Knowlton, Lisa Wyatt, and Cynthia C. Phillips, *The Logic Model Guidebook: Better Strategies for Great Results*, Thousand Oaks, Calif.: SAGE, 2009.

Larson, Meredith, "Career and Technical Education at IES," *Inside IES Research*, February 1, 2018. As of April 18, 2019:
https://ies.ed.gov/blogs/research/post/career-and-technical-education-at-ies

Lindholm-Leary, Kathryn, and Gary Hargett, *Evaluator's Toolkit for Dual Language Programs*, Sacramento, Calif.: California Department of Education, December 2006. As of April 15, 2019:
http://www.cal.org/twi/EvalToolkit/index.htm

Millenky, Megan, Dan Bloom, and Colleen Dillon, *Making the Transition: Interim Results of the National Guard Youth ChalleNGe Evaluation*, New York: MDRC, May 2010.

Millenky, Megan, Dan Bloom, Sara Muller-Ravett, and Joseph Broadus, *Staying on Course: Three-Year Results of the National Guard Youth ChalleNGe Evaluation*, New York: MDRC, June 2011.

National Guard Youth ChalleNGe, homepage, undated. As of October 17, 2017:
https://www.jointservicessupport.org/NGYCP/

———, *An Assessment of the Youth ChalleNGe Program: Performance and Accountability Highlights*, Arlington, Va.: National Guard Bureau, 2010.

———, *2015 Performance and Accountability Highlights*, Arlington, Va.: National Guard Bureau, December 2015. As of April 15, 2019:
https://prhome.defense.gov/Portals/52/Documents/MRA_Docs/RI/2015%20NGYCP%20Annual%20Report%20Final.pdf?ver=2018-01-30-185857-893

National Reporting System for Adult Education, "NRS Tips: Sampling for the Follow-Up Surveys," undated.

National Research Council, *The Importance of Common Metrics for Advancing Social Science Theory and Research: A Workshop Summary*, Washington, D.C.: National Academies Press, 2011.

Olsen, Marty, *Guide to Administering TABE (Tests of Adult Basic Education): A Handbook for Teachers and Test Administrators*, Little Rock, Ark.: Arkansas Department of Career Education, 2009.

Perez-Arce, Francisco, Louay Constant, David S. Loughran, and Lynn A. Karoly, *A Cost-Benefit Analysis of the National Guard Youth ChalleNGe Program,* Santa Monica, Calif.: RAND Corporation, TR-1193-NGYF, 2012. As of October 17, 2017:
http://www.rand.org/pubs/technical_reports/TR1193.html

Price, Hugh, "Foundations, Innovation and Social Change: A Quixotic Journey Turned Case Study," working paper prepared during practitioner residency, Bellagio, Italy: Rockefeller Foundation Bellagio Center, July 2010. As of October 17, 2017:
http://cspcs.sanford.duke.edu/sites/default/files/Foundations%20Innovation%20and%20Social%20Change.pdf

Schwartz, Sarah E. O., Jean E. Rhodes, Renée Spencer, and Jean B. Grossman, "Youth Initiated Mentoring: Investigating a New Approach to Working with Vulnerable Adolescents," *American Journal of Community Psychology*, Vol. 52, No. 1–2, 2013, pp. 155–169.

Shakman, Karen, and Sheila M. Rodriguez, *Logic Models for Program Design, Implementation, and Evaluation: Workshop Toolkit*, Washington, D.C.: Regional Educational Laboratory Northeast and Islands, May 2015. As of October 17, 2017:
http://files.eric.ed.gov/fulltext/ED556231.pdf

Spencer, Renée, Toni Tugenberg, Mia Ocean, Sarah E. O. Schwartz, and Jean E. Rhodes, "Somebody Who Was on My Side": A Qualitative Examination of Youth Initiated Mentoring, *Youth and Society,* Vol. 48, No. 3, 2016, pp. 402–424.

TABE—*See* Tests of Adult Basic Education.

Tests of Adult Basic Education, "States Using TABE," webpage, undated. As of December 7, 2017:
http://tabetest.com/resources-2/states-using-tabe/

United States Code, Title 32, Section 509, National Guard Youth ChalleNGe Program of Opportunities for Civilian Youth, January 3, 2012.

U.S. Department of Defense Instruction 1025.8, National Guard ChalleNGe Program, Washington, D.C.: U.S. Department of Defense, March 20, 2002.

U.S. Department of Education, Office of Vocational and Adult Education, Division of Adult Education and Literacy, *Implementation Guidelines: Measures and Methods for the National Reporting System for Adult Education*, Washington, D.C., February 2016. As of April 15, 2019:
https://files.eric.ed.gov/fulltext/ED584415.pdf

West Virginia Department of Education, *Correlation Between Various Placement Instruments for Reading, Language/Writing, Mathematics, Elementary Algebra*, Charleston, W. Va., undated. As of November 8, 2017:
https://wvde.state.wv.us/abe/documents/CorrelationBetweenVariousPlacementInstruments.pdf

Wenger, Jennie W., Louay Constant, and Linda Cottrell, *National Guard Youth ChalleNGe: Program Progress in 2016–2017*, Santa Monica, Calif.: RAND Corporation, RR-2276-OSD, 2018. As of January 15, 2019:
https://www.rand.org/pubs/research_reports/RR2276.html

Wenger, Jennie W., Louay Constant, Linda Cottrell, Thomas E. Trail, Michael J. D. Vermeer, and Stephani L. Wrabel, *National Guard Youth ChalleNGe: Program Progress in 2015–2016*, Santa Monica, Calif.: RAND Corporation, RR-1848-OSD, 2017. As of November 7, 2017:
https://www.rand.org/pubs/research_reports/RR1848.html

Wenger, Jennie W., Cathleen McHugh, and Lynda Houck, *Attrition Rates and Performance of ChalleNGe Participants over Time*, Arlington, Va.: Center for Naval Analyses, CRM D0013758.A2/Final, April 2006.

Wenger, Jennie W., Cathleen McHugh, Seema Sayala, and Robert Shuford, *Variations in Participants and Policies Across ChalleNGe Programs,* Arlington, Va.: Center for Naval Analyses, CRM D0017743.A2/Final, April 2008.